Holt Mathematics

Chapter 13 Resource Book

HOLT, RINEHART AND WINSTON
A Harcourt Education Company
Orlando • Austin • New York • San Diego • London

Printed in the United States of America

ISBN 0-03-078404-2

5 170 09 08

CONTENTS

Blackline Masters

Parent Letter 1
Lesson 13-1 Practice A, B, C 3
Lesson 13-1 Reteach 6
Lesson 13-1 Challenge 8
Lesson 13-1 Problem Solving 9
Lesson 13-1 Reading Strategies 10
Lesson 13-1 Puzzles, Twisters & Teasers 11
Lesson 13-2 Practice A, B, C 12
Lesson 13-2 Reteach 15
Lesson 13-2 Challenge 17
Lesson 13-2 Problem Solving 18
Lesson 13-2 Reading Strategies 19
Lesson 13-2 Puzzles, Twisters & Teasers 20
Lesson 13-3 Practice A, B, C 21
Lesson 13-3 Reteach 24
Lesson 13-3 Challenge 25
Lesson 13-3 Problem Solving 26
Lesson 13-3 Reading Strategies 27
Lesson 13-3 Puzzles, Twisters & Teasers 28
Lesson 13-4 Practice A, B, C 29
Lesson 13-4 Reteach 32
Lesson 13-4 Challenge 34
Lesson 13-4 Problem Solving 35
Lesson 13-4 Reading Strategies 36
Lesson 13-4 Puzzles, Twisters, & Teasers 37
Lesson 13-5 Practice A, B, C 38
Lesson 13-5 Reteach 41
Lesson 13-5 Challenge 43
Lesson 13-5 Problem Solving 44
Lesson 13-5 Reading Strategies 45
Lesson 13-5 Puzzles, Twisters & Teasers 46
Lesson 13-6 Practice A, B, C 47
Lesson 13-6 Reteach 50
Lesson 13-6 Challenge 52
Lesson 13-6 Problem Solving 53
Lesson 13-6 Reading Strategies 54
Lesson 13-6 Puzzles, Twisters & Teasers 55
Lesson 13-7 Practice A, B, C 56
Lesson 13-7 Reteach 59
Lesson 13-7 Challenge 60
Lesson 13-7 Problem Solving 61
Lesson 13-7 Reading Strategies 62
Lesson 13-7 Puzzles, Twisters & Teasers 63
Answers to Blackline Masters 64

Date ___________

Dear Family,

In this chapter, your child will learn about sequences and functions, expanding the knowledge of algebra already learned.

A sequence is a list of related numbers called *terms.* In an arithmetic sequence, the difference between one term and the next is always the same. This difference is called the common difference. The common difference is added to each term to get the next term.

Joanne received 5000 bonus miles for joining a frequent flier program. Each time she flies to visit her grandparents, she earns 1250 miles. The number of miles Joanne has in her account is 6250 after 1 trip, 7500 after 2 trips, 8750 after 3 trips, and so on.

After 1 trip	After 2 trips	After 3 trips	After 4 trips
6250	7500	8750	10,000

difference 1250 ⟶ *difference* 1250 ⟶ *difference* 1250

Your child will learn how to determine if a sequence is arithmetic and how to find the common difference.

8, 13, 18, 23, 28, . . .
5 5 5 5 *Subtract each term from the term before it.*
This sequence could be arithmetic with a common difference of 5.

In a geometric sequence, the ratio of one term and the next is always the same. This ratio is called the common ratio. The common ratio is multiplied by each term to get the next term.

96, 48, 24, 12, 6, . . .
$\frac{1}{2}$ $\frac{1}{2}$ $\frac{1}{2}$ $\frac{1}{2}$ *Divide each term by the term before it.*
The sequence could be geometric with a common ratio of $\frac{1}{2}$.

5, 7, 9, 11, . . .
$\frac{7}{5}$ $\frac{9}{7}$ $\frac{11}{9}$ *Divide each term by the term before it.*
The sequence is not geometric.

Your child will be introduced to functions. A function is a rule that relates two quantities so that each input value corresponds to exactly one output value. The domain is the set of all possible input values, and the range is the set of all possible output values.

Functions can be represented in several ways, including tables, graphs, and equations. If the domain of a function has infinitely many values, it is impossible to represent them all in a table, but a table can be used to show some of the values and to help in creating a graph.

x	$y = x^2 + 1$	y
−2	$y = (-2)^2 + 1$	5
−1	$y = (-1)^2 + 1$	2
0	$y = (0)^2 + 1$	1
1	$y = 1^2 + 1$	2
2	$y = 2^2 + 1$	5

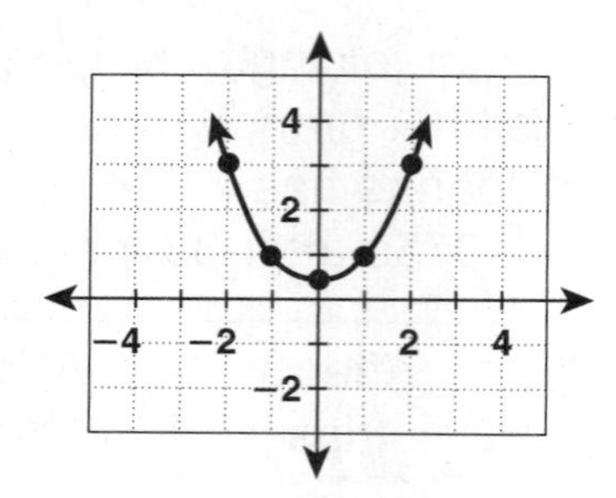

Your child will also learn about inverse variation.

Inverse Variation		
Words	**Numbers**	**Algebra**
An inverse variation is a relationship in which one variable quantity increases as another variable quantity decreases. The product of the variables is a constant.	$y = \frac{120}{x}$ $xy = 120$	$y = \frac{k}{x}$ $xy = k$

The table shows the number of days needed to construct a building for different sizes of work crew.

Crew Size	2	3	5	10	20
Days of Construction	90	60	36	18	9

$20(9) = 180$, $10(18) = 180$, $5(36) = 180$, $3(60) = 180$, $2(90) = 180$

$xy = 180$ *The product of x and y is always the same, 180.*

The relationship is an inverse variation: $y = \frac{180}{x}$.

For additional resources, visit g.hrw.com and enter the keyword MT7 Parent.

Name ______________________ Date __________ Class __________

LESSON **13-1**

Practice A

Terms of Arithmetic Sequences

Find the common difference for each arithmetic sequence.

1. 2, 4, 6, 8, 10, ...

____

2. 19, 13, 7, 1, −5, ...

____

3. −3, −6, −9, −12, −15, ...

____

4. −13, −9, −5, −1, 3, ...

____

5. 1.1, 2.2, 3.3, 4.4, 5.5, ...

____

6. 2, $\frac{3}{2}$, 1, $\frac{1}{2}$, 0, ...

____

Find the next three terms in each sequence.

7. 15, 11, 7, 3, −1, ...

____

8. −22, −28, −34, −40, −46, ...

____

9. −18, −13, −8, −3, 2, ...

____

10. 41, 32, 23, 14, 5, ...

____

11. −2.1, −3.7, −5.3, −6.9, −8.5, ...

____

12. $\frac{5}{3}$, $\frac{4}{3}$, 1, $\frac{2}{3}$, $\frac{1}{3}$, ...

____

Find the given term in each arithmetic sequence.

13. 10th term:
15, 24, 33, 42, 51, ...

____

14. 16th term:
5, 3, 1, −1, −3, ...

____

15. James is given 75 vocabulary words the first week in English class. He learns 10 words the first day and five more each day after that. How many days will it take James to learn all 75 vocabulary words?

____

Name ______________________ Date ____________ Class ____________

LESSON **13-1**

Practice B

Terms of Arithmetic Sequences

Determine if each sequence could be arithmetic. If so, give the common difference.

1. 18, 20, 22, 24, 26, ...

2. 48, 42, 36, 30, 24, ...

3. 15, 30, 60, 120, 240, ...

4. 10.4, 8.3, 6.2, 4.1, 2, ...

5. $\frac{1}{3}, \frac{1}{9}, \frac{1}{27}, \frac{1}{81}, \frac{1}{243}, \ldots$

6. 83, 66, 49, 32, 15, ...

7. 8.1, 2.7, 0.9, 0.3, 0.1, ...

8. $\frac{2}{3}, \frac{4}{3}, 2, \frac{8}{3}, \frac{10}{3}, \ldots$

9. −58, −35, −12, 11, 34, ...

Find the given term in each arithmetic sequence.

10. 14th term: 60, 68, 76, 84, 92, ...

11. 35th term: 3.5, 3.8, 4.1, 4.4, 4.7, ...

12. 21st term: 103, 84, 65, 46, 27, ...

13. 22nd term: −2, −5, −8, −11, −14, ...

14. 16th term: 73, 44, 15, −14, −43, ...

15. 50th term: −9, 2, 13, 24, 35, ...

16. 19th term: −87, −78, −69, −60, −51, ...

17. 25th term: $3\frac{1}{4}, 3\frac{1}{2}, 3\frac{3}{4}, 4, 4\frac{1}{4}, \ldots$

18. A cook started with 26 ounces of special sauce. She used 1.4 ounces of the sauce in each of a number of dishes and had 2.2 ounces left over. How many dishes did she make with the sauce?

19. Kuang started the basketball season with 54 points in his career. He scores 3 points more each game he plays. How many games will it take for him to have scored a total of 132 points in his basketball career?

Name ______________________ Date __________ Class __________

LESSON 13-1

Practice C

Terms of Arithmetic Sequences

Find the given term in each arithmetic sequence.

1. 14th term: 7, 2, −3, −8, −13, ...

____

2. 9th term: −12, −4, 4, 12, 20, ...

____

3. 20th term: $\frac{3}{4}, \frac{1}{4}, -\frac{1}{4}, -\frac{3}{4}, -1\frac{1}{4}, \ldots$

____

4. 11th term: 3.5, 5.2, 6.9, 8.6, 10.3, ...

____

Write the next three terms of each arithmetic sequence.

5. $\frac{3}{5}, \frac{6}{5}, \frac{9}{5}, \frac{12}{5}$, 3, ...

____

6. 3, −8, −19, −30, −41, ...

____

7. 8.9, 10.3, 11.7, 13.1, 14.5, ...

____

Write the first five terms of each arithmetic sequence.

8. $a_1 = 7,\ d = -5$

____

____

9. $a_1 = \frac{3}{4},\ d = -\frac{1}{2}$

____

____

10. $a_1 = 3.5,\ d = 1.7$

____

____

11. The 12th term of an arithmetic sequence is 100. The common difference is 8. What are the first five terms of the arithmetic sequence?

____

12. The 25th term of an arithmetic sequence is −157. The common difference is −6. What are the first five terms of the arithmetic sequence?

____

13. Neleh opens the bowling season with an average of 135. Each week she raises her average 2 points. At the end of the 20 week season, what will her average be?

____

Name ______________________ Date ___________ Class ___________

LESSON **13-1**

Reteach

Terms of Arithmetic Sequences

In an **arithmetic sequence,** the difference between terms is constant. The difference is called the **common difference.**

This is an arithmetic sequence with a common difference of 3.

2, 5, 8, 11, 14, ...

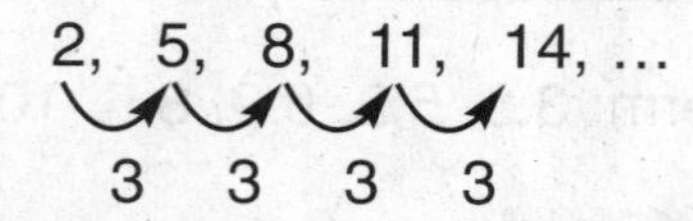

3 3 3 3

This is not an arithmetic sequence since there is no common difference.

2, 5, 9, 14, 20, ...

3 4 5 6

Complete to determine if each sequence is arithmetic.

1. 20, 16, 12, 8, 4, ...

−4 ____ ____ ____

arithmetic? __________

2. 1, 2, 4, 8, 16, ...

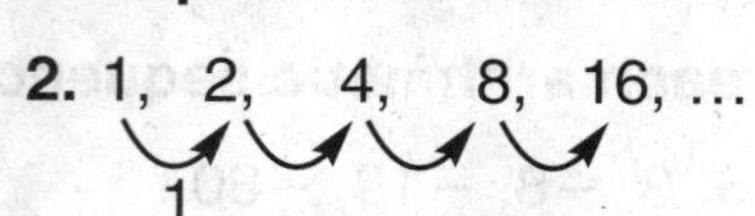

1 ____ ____ ____

arithmetic? __________

3. 0.1, 0.2, 0.3, 0.4, ...

____ ____ ____

arithmetic? __________

4. $\frac{1}{2}$, 1, $\frac{3}{2}$, 2, $\frac{5}{2}$, ...

____ ____ ____ ____

arithmetic? __________

5. 2, $\frac{3}{2}$, 1, $\frac{1}{2}$, 0, ...

____ ____ ____ ____

arithmetic? __________

6. 3, 1, 0, $-\frac{1}{2}$, $-\frac{1}{4}$, ...

____ ____ ____ ____

arithmetic? __________

You can use the common difference to find any term in an arithmetic sequence.

4, 6, 8, 10, 12, ... This arithmetic sequence has a common difference of 2.

2 2 2 2

This is the 1st term of the sequence.	4
For the 2nd term, add the common difference × 1	4 + 2 × 1 = 6
For the 3rd term, add the common difference × 2	4 + 2 × 2 = 8
For the 4th term, add the common difference × 3	4 + 2 × 3 = 10
For the 5th term, add the common difference × 4	4 + 2 × 4 = 12
For the *n*th term, add the common difference × (*n* − 1)	4 + 2 × (*n* − 1)

Complete to find the given term of the arithmetic sequence 4, 6, 8, 10, 12, ...

7. the 9th term

4 + 2 × _____ = _____

8. the 20th term

4 + 2 × _____ = _____

9. the 100th term

4 + 2 × _____ = _____

Name ______________________________ Date ____________ Class ____________

LESSON 13-1

Reteach

Terms of Arithmetic Sequences (continued)

You can use a formula to find the *n*th term, a_n, of an arithmetic sequence with common difference *d*. $\mathbf{a_n = a_1 + (n - 1)d}$

Find the 20th term of this arithmetic sequence: 2, 5, 8, 11, 14, ...

$a_n = a_1 + (n - 1)d$

$a_{20} = 2 + (20 - 1)3$

$a_{20} = 2 + (19)3 = 2 + 57 = 59$

Complete to find the given term of each arithmetic sequence.

10. 28, 34, 40, 46, 52, ...

Find the 18th term.

$n =$ _____, $a_1 =$ _____, $d =$ _____

$a_{18} =$ _____ + (_____ − 1)_____

= _____ + _____ = _____

11. $\frac{1}{8}, \frac{1}{4}, \frac{3}{8}, \frac{1}{2}, \frac{5}{8}, \ldots$

Find the 25th term.

$n =$ _____, $a_1 =$ _____, $d =$ _____

$a_{25} =$ _____ + (_____ − 1)_____

= _____ + _____ = _____

You can use the same formula to find other missing information.

What term of the arithmetic sequence 0.25, 0.50, 0.75, ... is 6.5?

Assign values to the variables.	$a_1 = 0.25, d = 0.25, a_n = 6.5, n = ?$
Substitute into the formula.	$a_n = a_1 + (n - 1)d$
Solve for *n*.	$6.5 = 0.25 + (n - 1)0.25$
Multiply.	$6.5 = 0.25 + 0.25n - 0.25$
Combine like terms.	$6.5 = 0.25n$
Divide.	$\frac{6.5}{0.25} = \frac{0.25n}{0.25}$
So, 6.5 is the 26th term.	$26 = n$

Complete to answer the question.

12. What term of the arithmetic sequence 8, 16, 24, 32, ... is 112?

$a_n = a_1 + (n - 1)d$

_____ = _____ + $(n - 1)$_____ Substitute.

________________________ Solve for *n*.

$a_1 =$ _____

$d =$ _____

____

$a_n =$ _____

____

So, 112 is the _____ term

_____ $= n$

Name ______________________ Date ____________ Class ____________

LESSON **13-1**

Challenge

Learn from "The Prince of Mathematics"

When the German mathematician Karl Gauss was a schoolboy, his teacher gave the problem of summing the integers from 1 through 100, hoping it would keep the class quiet. But young Karl wrote the correct answer after only a few seconds. How did he do it?

This is the sum. $S = 1 + 2 + 3 + \ldots + 98 + 99 + 100$

Reverse the numbers. $S = 100 + 99 + 98 + \ldots + 3 + 2 + 1$

Add vertically. $2S = 101 + 101 + 101 + \ldots + 101 + 101 + 101$

This sum contains one hundred addends of 101.

$$2S = 100(101)$$

$$\frac{2S}{2} = \frac{100(101)}{2}$$

$$S = \frac{100}{2}(101) = 50(101) = 5050$$

Gauss had come upon a method for finding the sum of any number of terms in an arithmetic sequence.

$$S_n = \frac{n}{2}(a_1 + a_n)$$

Applying the formula to the original problem:

Substitute $n = 100$, $a_1 = 1$, $a_n = 100$ $\quad S_{100} = \frac{100}{2}(1 + 100) = 50(101) = 5050$

1. Now that you know how to find the sum of the first 100 integers, and you know what the sum is, can you just divide by 2 to find the sum of the first 50 even integers? of the first 50 odd integers? Explain.

____

____

____

2. Find the sum of the first 750 integers.

____

3. Find the sum of the first 100 terms of this arithmetic sequence: 3, 6, 9, 12, …
(Hint: First find the 100th term.)

____

____

Name ______________________ Date ____________ Class ____________

LESSON 13-1

Problem Solving

Terms of Arithmetic Sequences

A section of seats in an auditorium has 18 seats in the first row. Each row has two more seats than the previous row. There are 25 rows in the section. Write the correct answer.

1. List the number of seats in the second, third and fourth rows of the section.

____

2. How many seats are in the 10th row?

____

3. How many seats are in the 15th row?

____

4. In which row are there 32 seats?

____

For 5–10, refer to the table below, which shows the boiling temperature of water at different altitudes. Choose the letter of the correct answer.

Altitude (thousands of feet)	**Boiling point of water** (°F)
1	210.2
2	208.4
3	206.6
4	204.8
5	203

5. What is the common difference?

A −1.8°F **C** −2.8°F
B 1.8°F **D** 6°F

6. According to the table, what would be the boiling point of water at an altitude of 10,000 feet?

F 192.2°F **H** 226.4°F
G 194°F **J** 228.2°F

7. According to the table, what would be the boiling point of water at an altitude of 15,000 feet?

A 181.4°F **C** 185°F
B 183.2°F **D** 235.4°F

8. Estimate the boiling point of water in Jacksonville, Florida, which has an elevation of 0 feet.

F 0°F **H** 212°F
G 208.4°F **J** 213.8°F

9. The highest point in the United States is Mt. McKinley, Alaska, with an elevation of 20,320 feet. Estimate the boiling point of water at the top of Mt. McKinley.

A 172.4°F **C** 244.4°F
B 176°F **D** 246.2°F

10. At which elevation will the boiling point of water be less than 150°F?

F 28,000 ft **H** 32,000 ft
G 30,000 ft **J** 35,000 ft

Name ______________________ Date ____________ Class ____________

LESSON 13-1

Reading Strategies

Focus On Vocabulary

A **sequence** is a list of numbers arranged in a certain order. Each number in a sequence is called a **term.**

An **arithmetic sequence** is a sequence in which the same number is added to each term to get the next term. The number that is added is called the **common difference.**

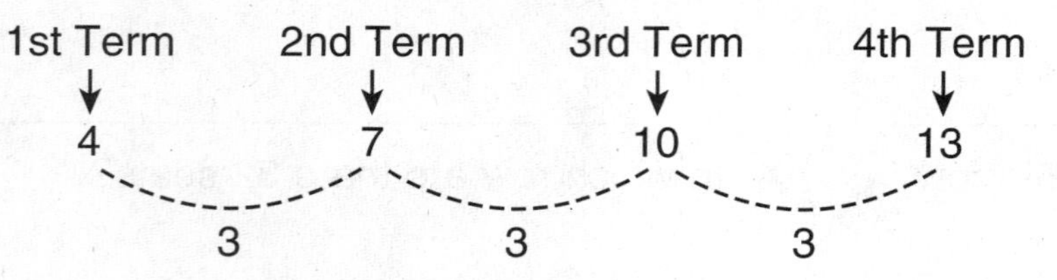

Answer the following questions.

1. What do you call the list of numbers above? ____________
2. What do you call each number in the list? ____________
3. How can you tell whether a sequence is an arithmetic sequence?

   ____

4. Is the list of numbers above an arithmetic sequence? Explain.

   ____

5. What is the common difference in the arithmetic sequence above? ____________
6. How do you find the common difference?

   ____

   ____

7. If the arithmetic sequence above continues, what number will come next in the sequence? ____________

Name ______________________ Date ____________ Class ____________

LESSON **13-1**

Puzzles, Twisters and Teasers

Take It or Leave It!

Find and circle words from the list in the word search (horizontally, vertically or diagonally). Find a word that answers the riddle. Circle it and write it on the line.

sequence	term	arithmetic	common	order
difference	finite	pattern	subscript	position

```
G V H B D I F F E R E N C E
S E Q U E N C E D S T Z X A
U J N H Y U O A Q W E C V R
B K J H G F M E R T R B N I
S B G T R W M Y U I M M L T
I U G O L V O R D E R F K H
P A T T E R N O P L K I J M
F O O T P R I N T S Q N H E
C D S U B S C R I P T I G T
N H Y T G B V F R E C T F I
P O S I T I O N A S D E D C
```

The more you take, the more you leave behind. What are they?

____

Name ______________________ Date ___________ Class ___________

LESSON 13-2

Practice A

Terms of Geometric Sequences

Find the common ratio for each of the following geometric sequences.

1. 5, 10, 20, 40, 80, ...

____

2. 3, 15, 75, 375, 1875, ...

____

3. 1, 6, 36, 216, 1296, ...

____

4. 0.5, 1.5, 4.5, 13.5, 40.5, ...

____

5. 810, 270, 90, 30, 10, ...

____

6. 72, 7.2, 0.72, 0.072, 0.0072, ...

____

Find the given term in each geometric sequence.

7. 7th term: 2, 6, 18, 54, ...

____

8. 10th term: 25, 5, 1, 0.2, ...

____

9. 8th term: 8, 4, 2, 1, ...

____

10. 6th term: 3, 4.5, 6.75, 10.125, ...

____

Find the next three terms of each geometric sequence.

11. 2, 10, 50, 250, 1250, ...

____

____

12. 4, 24, 144, 864, 5184, ...

____

____

13. 375, 75, 15, 3, 0.6, ...

____

____

14. $\frac{1}{3125}, \frac{1}{625}, \frac{1}{125}, \frac{1}{25}, \frac{1}{5}, \ldots$

____

15. 1.8, 3.6, 7.2, 14.4, 28.8, ...

____

16. 6804, 2268, 756, 252, 84, ...

____

17. Julie is doing an experiment. She is studying a cell that triples in number every hour. She started the experiment with 24 cells. How many cells are there at the end of 4 hours?

____

Name ______________________ Date ____________ Class ____________

LESSON **13-2**

Practice B

Terms of Geometric Sequences

Determine if each sequence could be geometric. If so, give the common ratio.

1. 4, 16, 64, 256, 1024, ...

____

2. 3, $\frac{3}{2}$, $\frac{3}{4}$, $\frac{3}{8}$, $\frac{3}{16}$, ...

____

3. 5, 10, 15, 20, 25, ...

____

4. 3, 18, 108, 648, 3888, ...

____

5. 1250, 125, 12.5, 1.25, 0.125, ...

____

6. 10, 15, 22.5, 33.75, 50.625, ...

____

7. 36, 12, 4, $\frac{4}{3}$, $\frac{4}{9}$, ...

____

8. 1440, 720, 240, 60, 12, ...

____

9. 9, 3, 1, 0.5, 0.25, ...

____

Find the given term in each geometric sequence.

10. 6th term: 25, 75, 225, 675, ...

____

11. 10th term: 320, 160, 80, 40, ...

____

12. 9th term: 4.5, 9, 18, 36, ...

____

13. 7th term: 0.02, 0.2, 2, 20, ...

____

14. 12th term: $\frac{1}{1000}$, $\frac{1}{100}$, $\frac{1}{10}$, 1, ...

____

15. 8th term: $\frac{3}{8}$, $\frac{3}{4}$, $\frac{3}{2}$, 3, ...

____

16. In an experiment a population of flies triples every week. The experiment starts with 12 flies. How many flies will there be by the end of week 5?

____

17. A small business earned $21 in its first month. It quadrupled this amount each month for the next several months. How much did the business earn in the 4th month?

____

Name ______________________ Date ____________ Class ____________

LESSON 13-2

Practice C

Terms of Geometric Sequences

Find the given term in each geometric sequence.

1. 8th term: $\frac{2}{81}, \frac{4}{27}, \frac{8}{9}, 5\frac{1}{3}, \ldots$ ____________

2. 7th term: 0.004, 0.04, 0.4, 4, ... ____________

Find the next three terms of each geometric sequence.

3. $a_1 = 3$, common ratio = 7 ____________

4. $a_1 = 800$, common ratio = 0.4 ____________

5. $a_1 = \frac{3}{8}$, common ratio = 2 ____________

6. $a_1 = 7.6$, common ratio = 8 ____________

Find the first five terms of each geometric sequence.

7. $a_1 = 250$, $r = 0.6$ ____________

8. $a_1 = 0.16$, $r = 5$ ____________

9. $a_1 = \frac{2}{81}$, $r = 6$ ____________

10. $a_1 = 0.004$, $r = 10$ ____________

11. $a_1 = 12$, $r = 9$ ____________

12. $a_1 = 320$, $r = \frac{1}{4}$ ____________

13. Find the 1st term of a geometric sequence with 5th term 11.664 and common ratio 0.6. ____________

14. Find the 1st and 7th terms of a geometric sequence with 3rd term $\frac{4}{9}$ and 4th term $\frac{8}{27}$.

____

15. Find the 1st term of a geometric sequence with 10th term −1024 and common ratio −2. ____________

16. Water is leaking from a water tower. On the first day two gallons were lost. The leak is getting progressively worse and the amount of water lost triples each day. How many gallons would be lost on the 8th day? ____________

Name ______________________ Date ____________ Class ____________

LESSON 13-2

Reteach

Terms of Geometric Sequences

In a **geometric sequence,** the ratio of one term to the next is constant. The ratio is called **common ratio**.

This is a geometric sequence with a common ratio of $\frac{3}{1} = 3$.

2, 6, 18, 54, 162, …

$\frac{3}{1}$ $\frac{3}{1}$ $\frac{3}{1}$ $\frac{3}{1}$

This is not a geometric sequence since there is no common ratio.

2, 4, 2, 4, 2, …

$\frac{2}{1}$ $\frac{1}{2}$ $\frac{2}{1}$ $\frac{1}{2}$

Complete to determine if each sequence is geometric. Write *yes* or *no*.

1. $\frac{1}{125}, \frac{1}{25}, \frac{1}{5}, 1, 5, \ldots$

$\frac{5}{1}$ _____ _____ _____

geometric? ________

2. $\frac{1}{8}, \frac{1}{4}, \frac{3}{8}, \frac{1}{2}, \frac{5}{8}, \ldots$

$\frac{2}{1}$ _____ _____ _____

geometric? ________

3. $4, -2, 1, -\frac{1}{2}, \frac{1}{4} \ldots$

_____ _____ _____ _____

geometric? ________

You can use the common ratio to find any term in a geometric sequence.

2, 8, 32, 128, …This geometric sequence has a common ratio of $\frac{4}{1}$ or 4.

4 4 4 4

This is the 1st term of the sequence.	2
Multiply it by the common ratio to get the 2nd term.	$2 \times 4 = 8$
Multiply it by the square of the common ratio to get the 3rd term.	$2 \times 4^2 = 32$
Multiply it by the cube of the common ratio to get the 4th term.	$2 \times 4^3 = 128$
To get the nth term, multiply the 1st term by the common ratio raised to the $(n - 1)$ power.	$2 \times 4^{n-1}$

Complete to find the given term of the geometric sequence 2, 8, 32, 128, …

4. 5th term

$2 \times 4^{\_} =$ ________

5. 7th term

$2 \times 4^{\_} =$ ________

6. 6th term

$2 \times 4^{\_} =$ ________

7. 10th term

$2 \times$ ____ = ________

8. 8th term

$2 \times$ ____ = ________

9. 9th term

$2 \times$ ____ = ________

Name ______________________ Date __________ Class __________

LESSON **13-2**

Reteach

Terms of Geometric Sequences (continued)

You can use a formula to find the nth term, a_n, of a geometric sequence with common ratio r. $\quad a_n = a_1 \bullet r^{n-1}$

Find the 7th term of this geometric sequence: 3, 6, 12, 24, 48, …

$a_n = a_1 \bullet r^{n-1}$ Find r. $r = \frac{6}{3} = 2$

$a_7 = 3 \bullet 2^{7-1}$ Substitute $n = 7$, $a_1 = 3$, $r = 2$.

$a_7 = 3 \bullet 2^{7-1} = 3 \bullet 2^6 = 3 \bullet 64 = 192$ The 7th term is 192.

Find the given term of each geometric sequence.

10. 1, 10, 100, 1000, . . .
Find the 9th term.

$n = 9$, $a_1 = 1$, $r = \frac{10}{1} =$ ______

$a_9 = 1 \bullet$ ______$^{9-1}$ $= 1 \bullet$ ______

$a_9 = 1 \bullet$ ______________

$=$ ______________

11. 1.1, 1.21, 1.331, 1.4641, . . .
Find the 7th term.

$n = 7$, $a_1 = 1.1$, $r = \frac{1.21}{1.1} =$ ______

$a_7 = 1.1 \times$ ______ $= 1.1 \times$ ______

$a_7 = 1.1 \times$ ______________

$=$ ______________

Which sequence has the greater 20th term? by how much?

1000, 1050, 1100, 1150, …
arithmetic sequence, $d = 50$

$a_{20} = 1000 + (20 - 1)50$

$a_{20} = 1000 + (19)50 = 1000 + 950$

$a_{20} = 1950$

2, 4, 8, 16, …
geometric sequence, $r = 2$

$a_{20} = 2 \bullet 2^{20-1}$

$a_{20} = 2 \bullet 2^{19} = 2 \bullet 524{,}288 = 1{,}048{,}576$

$a_{20} = 1{,}048{,}576$

The 20th term of this geometric sequence is greater by 1,046,626.

Determine whether each sequence is arithmetic or geometric and find its 15th term.

12. 2, 1, $\frac{1}{2}$, $\frac{1}{4}$, . . .

sequence is ______________

____

____

$a_{15} =$ ______________

13. $\frac{5}{2}$, 3, $\frac{7}{2}$

sequence is ______________

____

____

______________ $a_{15} =$ ______

Name ______________________ Date __________ Class __________

LESSON 13-2

Challenge
What's That Sum?

You can use a formula to find the sum of n terms of a geometric sequence with common ratio r. $S_n = \dfrac{a_1 - a_1 r^n}{1 - r}$

Find the sum of the first 5 terms of the geometric sequence 5, 15, 45, ...

$S_n = \dfrac{a_1 - a_1 r^n}{1 - r}$ Find r. $r = \dfrac{15}{5} = 3$

$S_5 = \dfrac{5 - 5 \cdot 3^5}{1 - 3}$ Substitute $n = 5$, $a_1 = 5$, $r = 3$.

$S_5 = \dfrac{5 - 5 \cdot 243}{1 - 3} = \dfrac{5 - 1215}{1 - 3} = \dfrac{-1210}{-2} = 605$

So, the sum of the first 5 terms of the sequence is 605.

Check: $5 + 15 + 45 + 135 + 405 = 605$

Use the formula to find each sum. Check your work by adding the terms with a calculator.

1. 32, 16, 8, ...
Find the sum of the first 6 terms.

$S_6 =$ __________

Check:

____

____

2. −3, 15, −75, ...
Find the sum of the first 5 terms.

$S_5 =$ __________

Check:

____

____

Name ______________________ Date __________ Class __________

LESSON 13-2

Problem Solving

Terms of Geometric Sequences

For Exercises 1–2, determine if the sequence could be geometric. If so, find the common ratio. Write the correct answer.

1. A computer that was worth $1000 when purchased was worth $800 after six months, $640 after a year, $512 after 18 months, and $409.60 after two years.

____

2. A student works for a starting wage of $6.00 per hour. She is told that she can expect a $0.25 raise every six months.

____

3. A piece of paper that is 0.01 inches thick is folded in half repeatedly. If the paper were folded 6 times, how thick would the result be?

____

4. A vacuum pump removes one-half of the air in a container with each stroke. How much of the original air is left in the container after 8 strokes?

____

For exercises 5–8, assume that the cost of a college education increases an average of 5% per year. Choose the letter of the correct answer.

5. If the in-state tuition at the University of Florida is $2256 per year, what will the tuition be in 10 years?

   A $3174.24

   B $3333.14

   C $3499.80

   D $3674.79

6. If it costs $3046 per year for tuition for a Virginia resident at the University of Virginia now, how much will tuition be in 8 years?

   F $4183.26

   G $4286.03

   H $4500.33

   J $4725.35

7. If it costs $25,839 per year in tuition to attend Northwestern University now, how much will tuition be in 5 years?

   A $31,407.47

   B $32,977.84

   C $37,965.97

   D $42,483.72

8. If you start attending Northwestern University in 5 years and attend for 4 years, how much will you spend in total for tuition?

   F $142,138.61

   G $135,370.12

   H $131,911.36

   J $169,934.88

Name ______________________ Date ____________ Class ____________

LESSON 13-2

Reading Strategies

Analyze Information

A **geometric sequence** is formed by *multiplication*. The same number is multiplied by each term in the sequence to get the next term. That number is called a **constant.**

Term	1st	2nd	3rd	4th
Value	2	6	18	54

Use the table to answer each question.

1. What is the 1st term in this geometric sequence? ____________

2. What is the 2nd term in this geometric sequence? ____________

3. What number do you multiply 2 by to get 6? ____________

4. What is the 3rd term in this geometric sequence? ____________

5. What number do you multiply 6 by to get 18? ____________

6. What is the constant in this geometric sequence? ____________

Multiplying by a constant in a geometric sequence results in a **common ratio** from one term to the next.

Use the table above to answer each question.

7. What is the ratio of the 2nd term to the 1st term? ____________

8. What is the ratio of the 3rd term to the 2nd term? ____________

9. What is the common ratio of this geometric sequence? ____________

Name ________________ Date __________ Class __________

LESSON 13-2

Puzzles, Twisters and Teasers

Itching for Answers!

Find the given term in each sequence. Use the answers to solve the riddle.

1. 12th term: 3, 6, 12, 24, 48 ... __________ T

2. 8th term: 1, 4, 16, 64, 256 ... __________ S

3. 9th term: −4, −2, 0, 2, 4 ... __________ F

4. 6th term: $\frac{1}{2}$, 1, 2, 4 ... __________ E

5. 7th term: 3, 6, 12, 24 ... __________ L

6. 5th term: 1, 1.5, 2.25, 3.375 ... __________ C

7. 10th term: $\frac{1}{3}, \frac{-1}{3}, \frac{1}{3}, \frac{-1}{3}, \frac{1}{3}$ ... __________ K

8. 6th term: 96, 48, 24, 12, 6 ... __________ A

9. 10th term: 5, −5, 5, −5, 5 ... __________ R

10. 6th term: 1, 2, 4, 8 ... __________ M

Where should you never take a dog?

To a ___ ___ ___ ___ ___ ___ ___ ___ ___ ___

12 192 16 3 32 3 −5 $-\frac{1}{3}$ 16 6144

Name ______________________ Date ____________ Class ____________

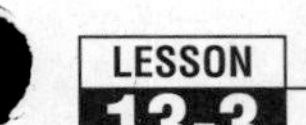

Practice A

Other Sequences

Complete the table by finding the first and second differences in each sequence.

1.

Sequence	2	5	11	20	32	47	65
1st Differences							
2nd Differences							

2.

Sequence	5	12	21	32	45	60	77
1st Differences							
2nd Differences							

Use first and second differences to find the next three terms in each sequence.

3. 3, 8, 18, 33, 53, …

____

4. 5, 6, 8, 12, 19, …

____

5. 6, 8, 12, 18, 26, …

____

6. 1.5, 2.5, 5.5, 10.5, 17.5, …

____

7. 7, 9, 13, 20, 31, …

____

8. $\frac{1}{2}$, 1, 2, $3\frac{1}{2}$, $5\frac{1}{2}$, …

____

Find the first five terms of each sequence defined by the given rule.

9. $a_n = 2n + 6$

____

10. $a_n = \frac{n-2}{n}$

____

11. $a_n = \frac{3n-1}{2}$

____

12. The rule of a sequence is to square the number of each term's position and add 2. Find the first six terms of the sequence.

____

Name ______________________ Date __________ Class __________

LESSON 13-3

Practice B

Other Sequences

Use first and second differences to find the next three terms in each sequence.

1. 3, 6, 10, 15, 21, ...

____

2. 11, 14, 18, 25, 37, ...

____

3. 10, 16, $22\frac{1}{3}$, 29, 36, ...

____

4. 14.5, 22.5, 31, 40, 49.5, ...

____

Give the next three terms in each sequence using the simplest rule you can find.

5. 6, 7, 10, 19, 38, ...

____

6. 0.5, 2, 4.5, 8, 12.5, ...

____

7. 36, 55, 80, 111, 148, ...

____

8. 3, 10, 21, 36, 55, ...

____

9. 1, 6, 15, 28, 45, ...

____

10. 0, 11, 30, 57, 92, ...

____

Find the first five terms of each sequence defined by the given rule.

11. $a_n = \frac{n^2 + 2}{n}$

____

12. $a_n = \frac{5n - 2}{n + 1}$

____

13. $a_n = \frac{3n^2}{n + 2}$

____

14. Suppose *a*, *b*, and *c* are three consecutive numbers in the Fibonacci sequence. Complete the following table and guess the pattern.

a, b, c	*ab*	*bc*
1, 1, 2		
2, 3, 5		
5, 8, 13		
13, 21, 34		
34, 55, 89		

____

Name ______________________ Date ____________ Class ____________

LESSON **13-3**

Practice C
Other Sequences

Give the next three terms in each sequence using the simplest rule you can find.

1. 14, 25, 36, 47, 58, ...

____

2. 9, 12, 17, 24, 33, ...

____

3. $1\frac{1}{5}$, $4\frac{4}{5}$, $10\frac{4}{5}$, $19\frac{1}{5}$, 30, ...

____

4. 10.5, 84, 283.5, 672, 1312.5, ...

____

Find the first five terms of each sequence defined by the given rule.

5. $a_n = 2n^2 + 6$

____

6. $a_n = \frac{12n - 5}{n}$

____

7. $a_n = \frac{n^2 - n}{2n}$

____

8. $a_n = \frac{4n^2 + 2n}{n + 2}$

____

____

9. $a_n = \frac{3n^2 - 2n}{2n + 1}$

____

____

10. $a_n = \frac{6n^2 - 2n + 3}{n + 2}$

____

____

11. Suppose *a, b, c, d, e, f, g, h, i,* and *j* are ten consecutive numbers in the Fibonacci sequence. Complete the following table and guess the pattern.

a, b, c, d, e, f, g, h, i, j	*11 • g*	*a + b + c + d + e + f + g + h + i + j*
1, 1, 2, 3,5, 8, 13, 21, 34, 55		
55, 89, 144, 233, 377, 610, 987, 1597, 2584, 4181		

____

____

12. The sum of any 10 consecutive terms in the Fibonacci sequence is divisible by what number?

____

Name ______________________ Date ____________ Class ____________

LESSON 13-3

Reteach

Other Sequences

Differences can help you find patterns in some sequences.

Find the next number in the sequence: 1, 6, 15, 28, 45, ...

1, 6, 15, 28, 45, **66** ...

Find the **first differences.** 5 9 13 17 **21**

Find the **second differences.** 4 4 4 4

Use the second and first diffences to calculate the next term.

Complete to find the next term in each sequence.

1. 1, 4, 9, 16, 25, ...

3 5 7 9

2 2 2

The next term in the sequence is:

25 + ______ = ______

2. 1, 8, 21, 40, 65, ...

7 13 19 25

6 6 6

The next term in the sequence is:

65 + ______ = ______

A rule is used to define a sequence.

Write a rule for this sequence: $\frac{1}{2}, \frac{2}{3}, \frac{3}{4}, \frac{4}{5}, \frac{5}{6}, \ldots$

A possible rule is that the numerator of a term is the number of that terms position, and the denominator is 1 more than the numerator.

This can be written algebraically as $a_n = \frac{n}{n+1}$.

Using this rule, the 10th term of the sequence is $a_{10} = \frac{10}{10+1} = \frac{10}{11}$.

Use the given rule to write the 5th and 10th terms.

3. 1, 8, 27, 64, ...

$a_n = n^3$

$a_5 = (\_\_\_\_\_\_)^3 = \_\_\_\_\_\_$

$a_{10} = (\_\_\_\_\_\_)^3 = \_\_\_\_\_\_$

4. 1, 3, 6, 10, ...

$a_n = \frac{n(n+1)}{2}$

$a_5 = \frac{(\ \ )(\ \ +1)}{2} = \_\_\_\_\_\_$

$a_{10} = \frac{(\ \ )(\ \ +1)}{2} = \_\_\_\_\_\_$

5. 1, −3, 1, −3, ...

$a_n = 2(-1)^{n+1} - 1$

$a_5 = 2(-1)^{\_\_ + 1} - 1$

$= \_\_\_\_\_\_$

$a_{10} = 2(-1)^{\_\_ + 1} - 1$

$= \_\_\_\_\_\_$

Name ______________________ Date __________ Class __________

LESSON 13-3

Challenge

Follow the Short-Cut

The Greek capital letter *sigma*, Σ, is used to mean "take the sum of what follows."

What follows Σ is a general term that is a rule for each term of a *summation*.

The general term is written with an *index*, shown by one letter (often, n).

The *limits* for the index are written above Σ (upper limit) and below Σ (lower limit).

The first term of the summation is formed by substituting the lower limit for the index into the general term.

Each succeeding term of the summation is formed by using successive integral values of the index, until the upper limit is reached.

$\sum_{n=1}^{5} 2^n$ This notation means take the sum of terms of the form 2^n for consecutive integral values of n beginning with $n = 1$ and ending with $n = 5$.

$$\sum_{n=1}^{5} 2^n = 2^1 + 2^2 + 2^3 + 2^4 + 2^5$$

$$\sum_{n=1}^{5} 2^n = 2 + 4 + 8 + 16 + 32 = 62$$

Evaluate each summation.

1. $\sum_{n=1}^{5} 4n$

= ______________________

= ______________________

= ______________________

2. $\sum_{n=1}^{3} (n^2 + 1)$

= ______________________

= ______________________

= ______________________

3. $\sum_{n=1}^{4} \frac{6}{n}$

= ______________________

= ______________

= ____

4. $\sum_{n=1}^{4} \frac{n}{n+1}$

= ______________________

____

= ______________________

= ______________________

Name ______________________ Date __________ Class __________

LESSON 13-3

Problem Solving

Other Sequences

A toy rocket is launched and the height of the rocket during its first four seconds is recorded. Write the correct answer.

Time (sec)	**Height** (ft)
0	0
1	176
2	320
3	432
4	512
5	
6	
7	

1. Find the first differences for the rocket's heights.

____

2. Find the second differences.

____

3. Use the first and second differences to predict the height of the rocket at 5, 6, and 7 seconds.

4. What is the maximum height of the rocket?

____

5. When will the rocket hit the ground?

____

For exercises 6–9, refer to the table below, which shows the number of diagonals for different polygons. Choose the letter for the correct answer.

Polygon	**Sides**	**Diagonals**
Triangle	3	0
Quadrilateral	4	2
Pentagon	5	5
Hexagon	6	9
Heptagon	7	14

6. What are the first differences for the diagonals?

A 1, 1, 1, 1
B 3, 2, 0, 3, 7
C 2, 3, 4, 5
D 2, 7, 14, 23

7. What are the second differences?

F 1, 1, 1
G 1, 2, 3, 4
H 5, 7, 8
J 7, 9, 11, 13

8. How many diagonals does a nonagon (9 sides) have?

A 21
B 24
C 27
D 32

9. Which rule will give the number of diagonals d for s sides?

F $d = \frac{s(s+1)}{2}$

G $d = (s-3)(s-2) - 1$

H $d = \frac{s(s-3)}{2}$

J $d = (s-3)(s-2)$

Name ______________________ Date ____________ Class ____________

LESSON 13-3

Reading Strategies

Look for a Pattern

To continue a sequence or find the rule for the sequence, **look for a pattern.** You can analyze increases (or decreases) from one term to the next to find the pattern.

Term	1st	2nd	3rd	4th	5th
Value	2	5	9	14	20

3 4

Use the table to answer each question.

1. What is the 1st term in the sequence? ____________
2. What is the 2nd term of the sequence? ____________
3. What is the increase in value from the 1st term to the 2nd term? ____________
4. What is the increase in value between the 2nd and 3rd terms? ____________
5. What is the increase between the 3rd and 4th terms? ____________
6. List in order the increases you found between one number and the next in the sequence above. ____________
7. What pattern do you see in the increases?

____

____

8. What will the 6th term in the sequence be? ____________
9. What will the 10th term in the sequence be? ____________

Name ______________________ Date __________ Class __________

LESSON 13-3

Puzzles, Twisters and Teasers

Puzzle Pattern!

Across

1. In a __________ sequence, add the two previous terms to find the next term.

3. If you do not see a pattern in the first diffrences, try finding the __________ differences.

5. To continue a sequence, look for a __________.

7. Some sequences are defined by a given __________.

8. First and second differences can help you find patterns in some __________.

9. When looking for a sequence with no given rule, try the __________ rule first.

Down

2. Sometimes a(n) __________ rule is used to define a sequence.

4. To begin, look for a pattern using first __________.

6. In a sequence with no given rule, you cannot __________ what the next term will be.

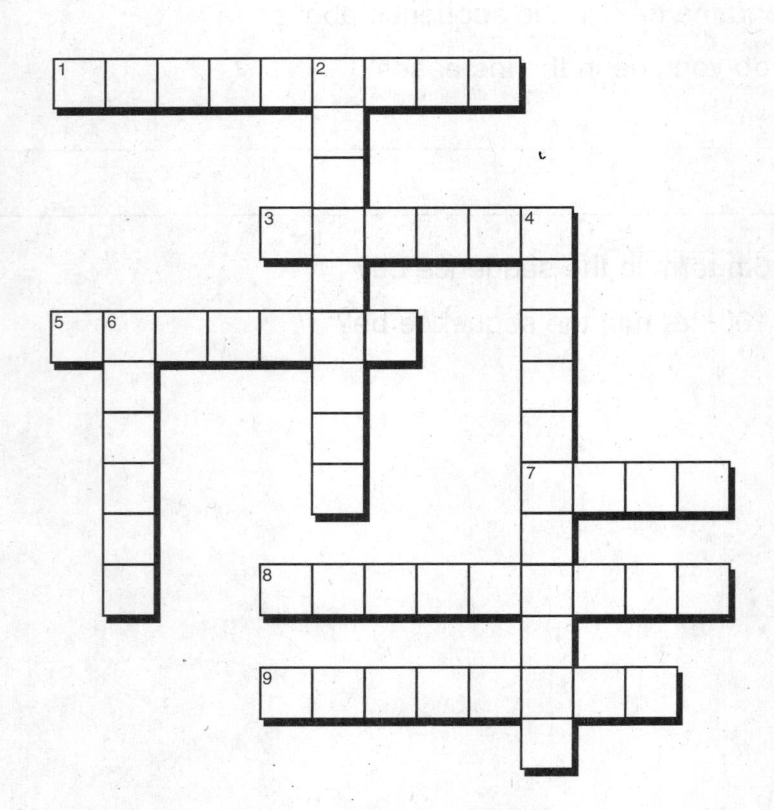

Name ________________________ Date ____________ Class ____________

LESSON 13-4

Practice A

Linear Functions

Graph each linear and write a rule for the function.

1.

x	y
−2	0
−1	−1
0	−2
1	−3
2	−4

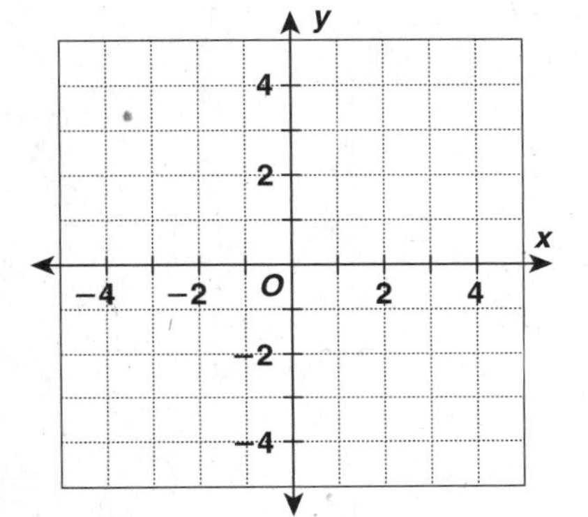

____

2.

x	y
−2	4
−1	2
0	0
1	−2
2	−4

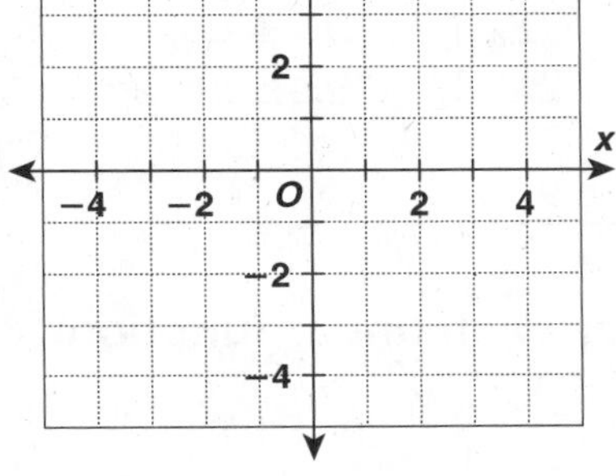

____

Write the rule for each linear function.

3.

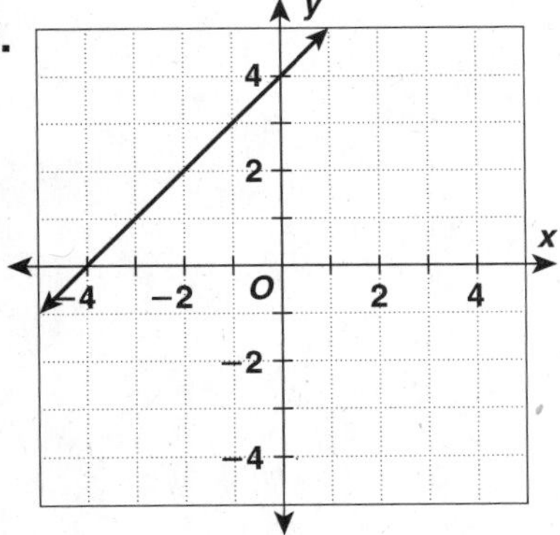

____

4.

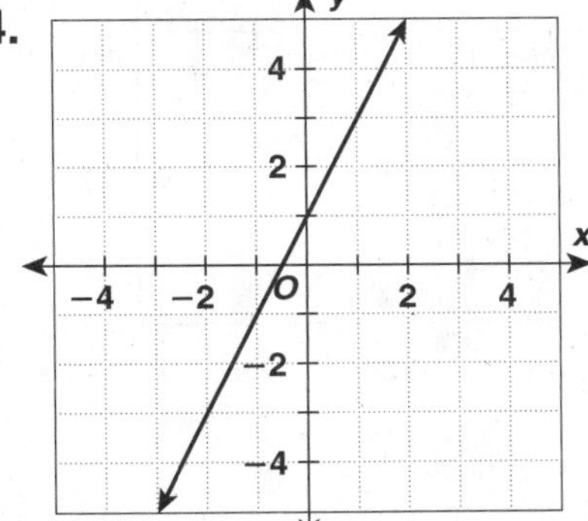

____

5. A salesperson receives a base monthly salary of $400, plus 5% of her total sales for the month. Find a rule for the linear function that describes her monthly salary. Use it to determine her salary if her total sales in January are $22,400.

____

____

Name ______________________ Date ____________ Class ____________

LESSON 13-4

Practice B

Linear Functions

Determine whether each function is linear.

1. $f(x) = -3x + 2$

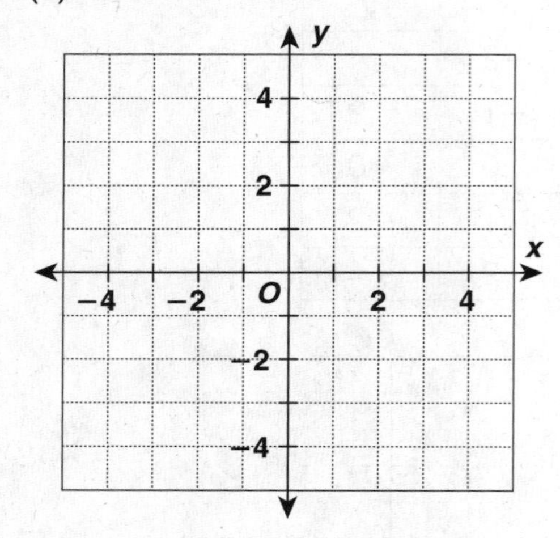

____

2. $f(x) = x^2 - 1$

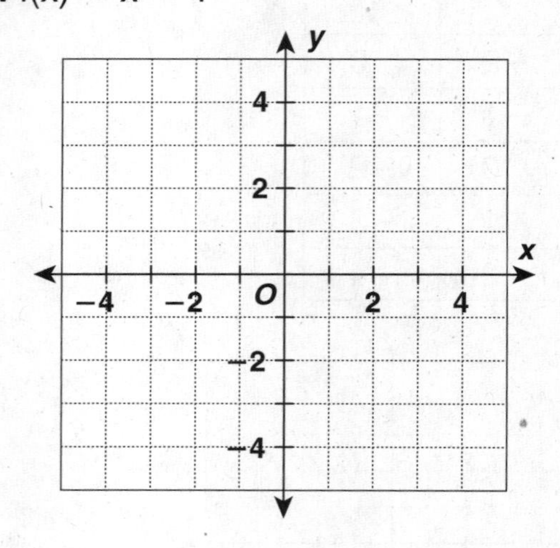

____

Write a rule for each linear function.

3.

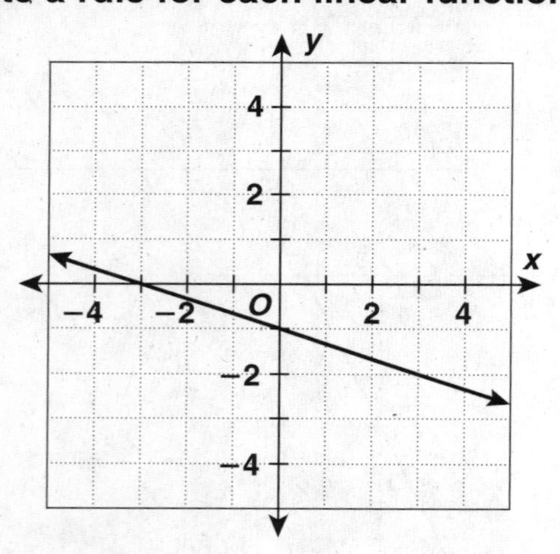

____

4.

x	y
−3	16
−1	12
3	4
7	−4

____

5. At the Sweater Store, the price of a sweater is 20% more than the wholesale cost, plus a markup of $8. Find a rule for a linear function that describes the price of sweaters at the Sweater Store. Use it to determine the price of a sweater with a wholesale cost of $24.50.

____

____

Name ______________________ Date __________ Class __________

LESSON 13-4

Practice C

Linear Functions

Write the rule for each linear function.

1.

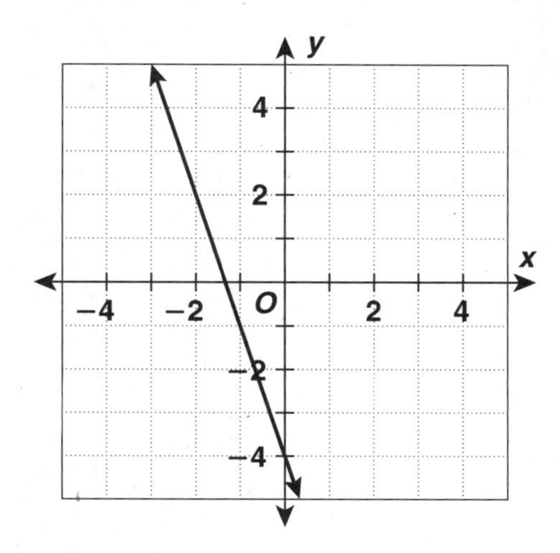

____

2.

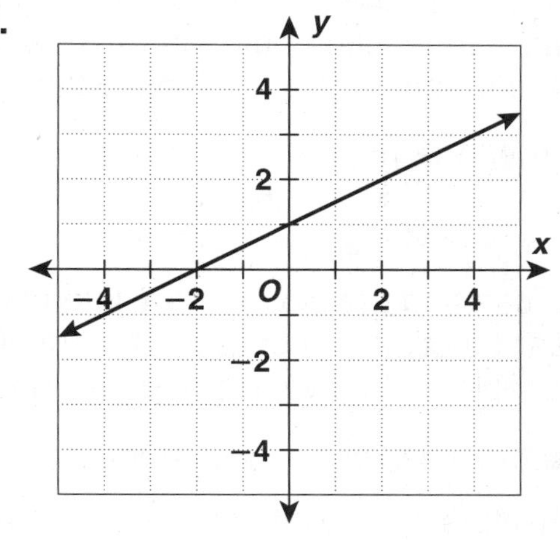

____

3.

x	y
−2	2.5
−1	1.5
0	0.5
2	−2.5

____

4.

x	y
−2	$-1\frac{1}{2}$
−1	$-1\frac{1}{4}$
0	−1
2	$-\frac{1}{2}$

____

5. A cab company charges its customers a flat rate of \$10 plus \$0.75 per mile. Find a rule for the linear function that describes the cab company's basic fees. ______________

6. Use the rule for Exercise 5 to find the cab company's charge for a cab ride of 24 miles. ______________

7. A company buys a forklift truck for \$10,000. The company depreciates the truck \$125 each month for x months. Write a linear function for the forklift's value after x months. ______________

8. Use the rule for Exercise 7 to find the depreciated value of the forklift truck after 15 months. ______________

Name ______________________________ Date __________ Class __________

LESSON 13-4

Reteach

Linear Functions

The graph of a **linear function** is a straight line, so you can write a **rule** for a linear function in slope-intercept form. Use function notation to show that the output value, $f(x)$, corresponds to the input value, x.

$$f(x) = mx + b$$

slope (m) y-intercept (b)

You can find the slope and y-intercept of a linear function in a graph of the function or in a table of its x-values and y-values.

$f(x) = 2x + 1$

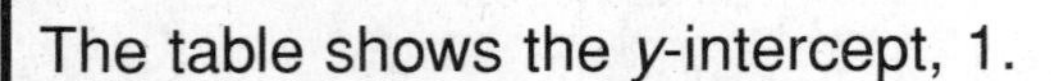

x	y
−1	−1
0	1
1	3
2	5

The table shows the y-intercept, 1. $b = 1$

Substitute 1 for b into the slope-intercept form. $f(x) = mx + 1$

Substitute a pair of x- and y-values and solve. $5 = m \cdot 2 + 1$

$4 = 2m$

$2 = m$

So, the rule for the function is $f(x) = 2x + 1$.

Write the rule for the linear function.

1.

x	y
−2	−1
−1	1
0	3
1	5

$f(x) = mx + b$

$f(x) = mx +$ ____

$(x, y) = (1, 5)$ $5 = m($____$) +$ ____

____ $= m$

$f(x) =$ ____________

2.

x	y
−1	3
0	−1
1	−5
2	−9

$f(x) = mx + b$

$f(x) = mx$ ______

$(x, y) = (-1, 3)$ ____ $= m($____$)$ ______

____ $= m$

$f(x) =$ __________

Name ______________________ Date __________ Class __________

LESSON 13-4

Reteach

Linear Functions (continued)

If a table does not contain the y-intercept, use two points to find the slope.

x	y
−2	−8
−1	−5
1	1
2	4

Use (1, 1) and (2, 4).

$m = \frac{4-1}{2-1} = \frac{3}{1} = 3$

Then substitute the value for m and the coordinates of any point, say (1, 1), and solve for b.

$f(x) = mx + b$
$1 = 3(1) + b$
$1 = 3 + b$
$-3 \quad -3$
$-2 = b$

Use the values of m and b to write the function rule: $f(x) = 3x - 2$

Write the rule for the linear function.

3.

x	y
−3	−17
−1	−7
1	3
3	13

Using (1, 3) and (3, 13);

$m =$ ________ = ________ = ____

Using your value for m and (1, 3):

$f(x) = mx + b$

____ = ____ + b

____ = b

$f(x) =$ ____________

Use the graph to write the function rule.

4. From point B, read the y-intercept.

$b =$ ____

Write the coordinates of two points on the line.

A(____), C(____)

Use these points to find the slope.

$m =$ ____________ = ______ = ____

$f(x) =$ ____________

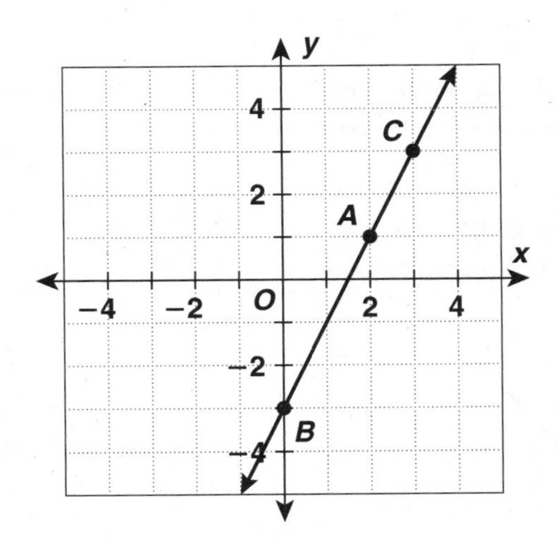

Name ______________________ Date __________ Class __________

LESSON 13-4

Challenge
Function? Absolutely!

Now, you will explore a unique function, the **absolute-value function $f(x) = |x|$.**

1. a. Complete this table of values and plot the points.

x	$\lvert x \rvert$	y
−3		3
−2		2
−1		1
0		0
1		1
2		2
3		3

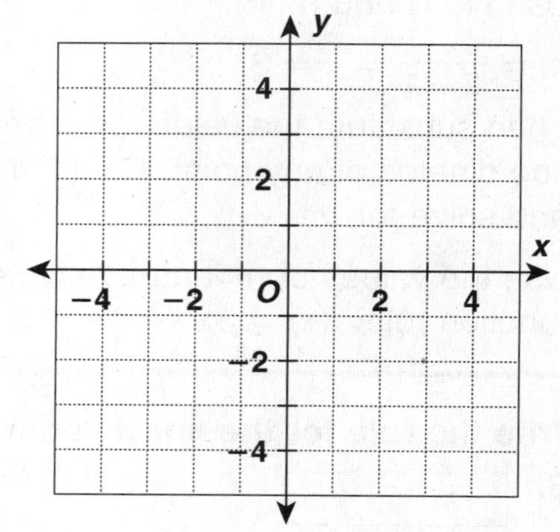

b. Your result should be in a V-formation. Connect the dots to show this.

c. Add more points to your graph that satisfy the absolute-value function $f(x) = |x|$.

d. domain? ______________ range? __________

If you have access to a graphing instrument (calculator or computer), use it for the following Exercises. If not, make tables.

2. a. On the same grid, draw the graphs of $f(x) = |x - 1|$ and $f(x) = |x - 2|$.

b. On the same grid, draw the graphs of $f(x) = |x + 1|$ and $f(x) = |x + 2|$.

c. What type of transformation takes the graph of $f(x) = |x|$ to the graphs you have drawn?

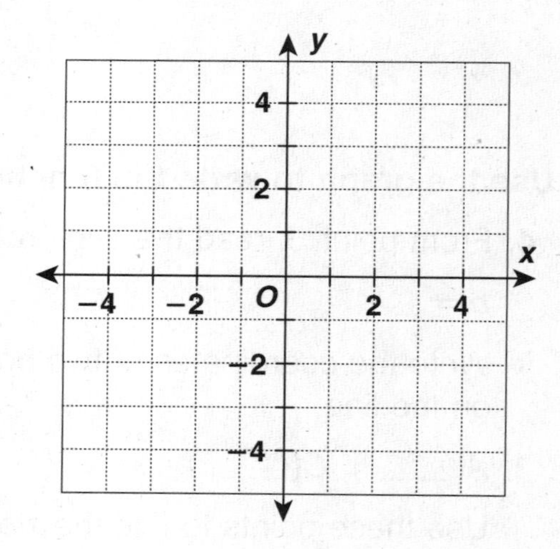

____

3. Predict the location of the vertex of the V-shape for each function that follows. (Write the coordinates.) Verify your predictions.

a. $f(x) = |x - 5|$ __________

b. $f(x) = |x + 5|$ __________

c. $f(x) = |x| + 1$ __________

d. $f(x) = |x| - 1$ __________

Name ______________________ Date __________ Class __________

LESSON 13-4

Problem Solving

Linear Functions

Write the correct answer.

1. The greatest amount of snow that has ever fallen in a 24-hour period in North America was on April 14–15, 1921 in Silver Lake, Colorado. In 24 hours, 76 inches of snow fell, at an average rate of 3.2 inches per hour. Find a rule for the linear function that describes the amount of snow after *x* hours at the average rate.

____

2. At the average rate of snowfall from Exercise 1, how much snow had fallen in 15 hours?

____

3. The altitude of clouds in feet can be found by multiplying the difference between the temperature and the dew point by 228. If the temperature is 75°, find a rule for the linear function that describes the height of the clouds with dew point *x*.

____

4. If the temperature is 75° and the dew point is 40°, what is the height of the clouds?

____

For exercises 5–7, refer to the table below, which shows the relationship between the number of times a cricket chirps in a minute and temperature.

Cricket Chirps/min	Temperature (°F)
80	60
100	65
120	70
140	75

5. Find a rule for the linear function that describes the temperature based on *x*, the number of cricket chirps in a minute based on temperature.

   A $f(x) = x + 5$

   B $f(x) = \frac{x}{4} + 40$

   C $f(x) = x - 20$

   D $f(x) = \frac{x}{2} + 20$

6. What is the temperature if a cricket chirps 150 times in a minute?

   F 77.5°F

   G 95°F

   H 130°F

   J 155°F

7. If the temperature is 85°F, how many times will a cricket chirp in a minute?

   A 61

   B 105

   C 180

   D 200

Name ______________________ Date __________ Class __________

LESSON 13-4

Reading Strategies

Use a Graphic Aid

The graph of a **linear function** is a straight line, so you can write a **rule** for a linear function in slope-intercept form.

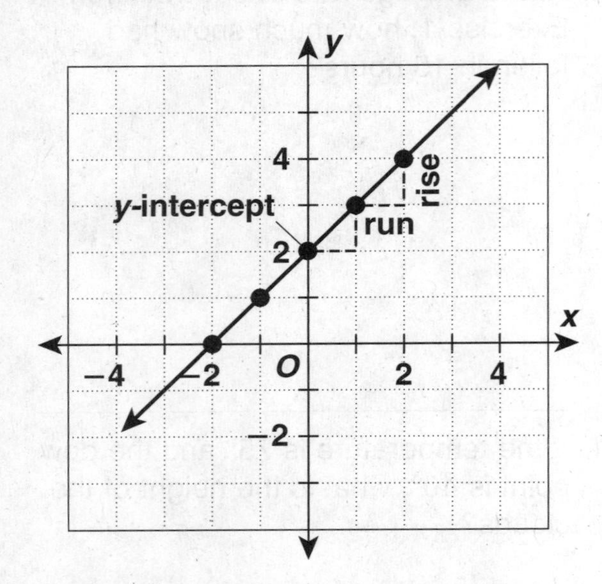

$f(x) = mx + b$

slope   intercept

Use the graph to answer each question.

1. What type of function is pictured in the graph? __________
2. What is the name for the y-coordinate of the point where the line crosses the *y*-axis? __________
3. What is the *y*-intercept for this graph? __________

You can use the graph to find the slope of the line. Find the "rise over run," or the change in *y* to the change in *x*.

$$\textbf{slope} = \frac{\textbf{rise}}{\textbf{run}} = \frac{\textbf{change in } y}{\textbf{change in } x}$$

Answer each question.

4. How can you find the slope of a linear function?

____

____

5. Find the slope of the graph above. __________
6. Complete to write a rule for the function. $f(x) =$ ______ $x +$ ______

Name ______________________ Date ____________ Class ____________

LESSON 13-4

Puzzles, Twisters and Teasers

Bee-lieve It or Not!

Fill in the missing coordinate for each point on the graph. Use the corresponding letter to solve the riddle. (Hint: There is one letter that you will not need.)

(−7, ____)	**E**	(1, ____)	**X**
(−6, ____)	**S**	(3, ____)	**G**
(−5, ____)	**B**	(5, ____)	**L**
(−3, ____)	**P**	(6, ____)	**N**
(−1, ____)	**A**	(7, ____)	**I**

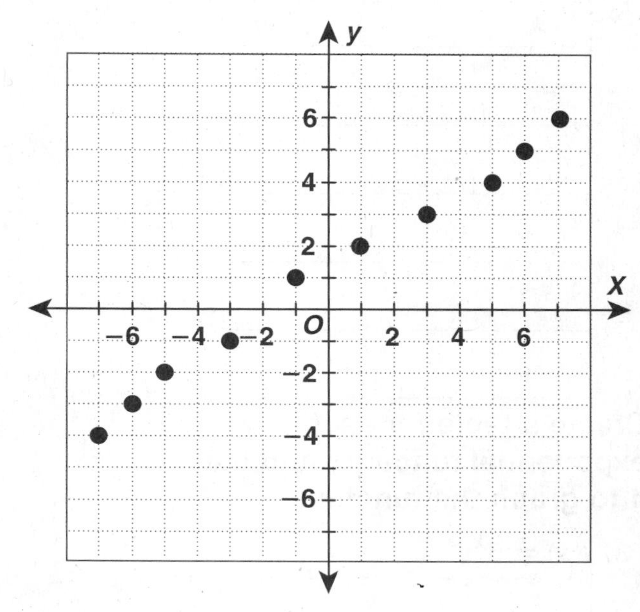

What is more amazing than a talking dog?

____
1

____ ____ ____ ____ L ____ ____ ____
−3 −1 −4 4 6 5 3

____ E ____
−2 −4

Name ______________________ Date __________ Class __________

LESSON 13-5

Practice A

Exponential Functions

Complete the table for each exponential functions.

1. $f(x) = 2^x$

x	y
−2	$y = 2^{-2} = \frac{1}{4}$
−1	
0	
1	
2	

2. $f(x) = (0.2)4^x$

x	y
−2	
−1	
0	
1	
2	

Create a table for each exponential function, and use it to graph the function.

3. $f(x) = 5^x$

x	y
−2	
−1	
0	
1	

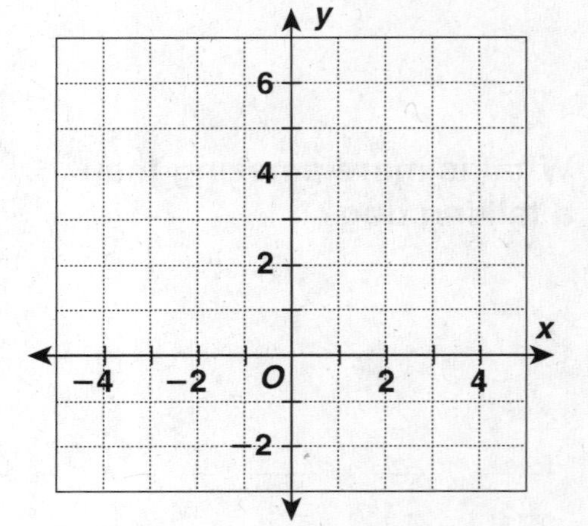

4. $f(x) = 4 \bullet 2^x$

x	y
−2	
−1	
0	
1	

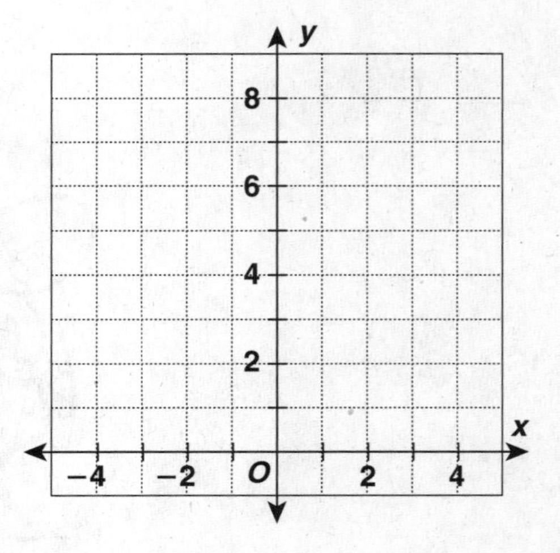

5. The exponential function $f(x) = 1000 \bullet 1.03^x$ describes the increase in a bank deposit of $1000 with a 3% annual interest rate for x years. Find the value of the deposit after 2 years.

____

Name ______________________ Date ____________ Class ____________

LESSON 13-5

Practice B

Exponential Functions

Create a table for each exponential function, and use it to graph the function.

1. $f(x) = 0.5 \cdot 4^x$

x	y
−1	$y = 0.5 \cdot 4^{-1} = 0.125$
0	
1	
2	

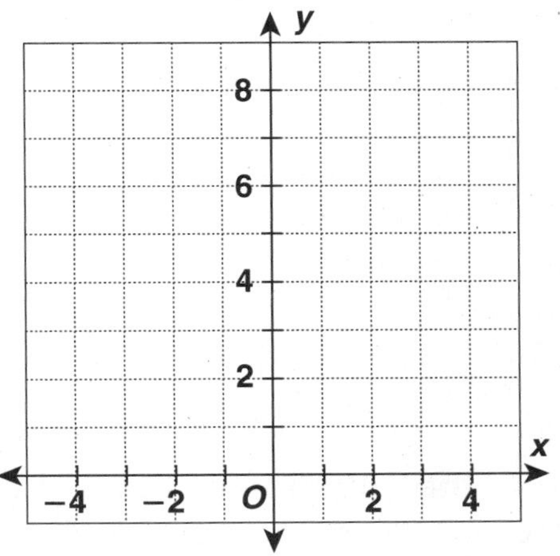

2. $f(x) = \frac{1}{3} \cdot 3^x$

x	y
−1	$y = \frac{1}{3} \cdot 3^{-1} = \frac{1}{9}$
0	
1	
2	

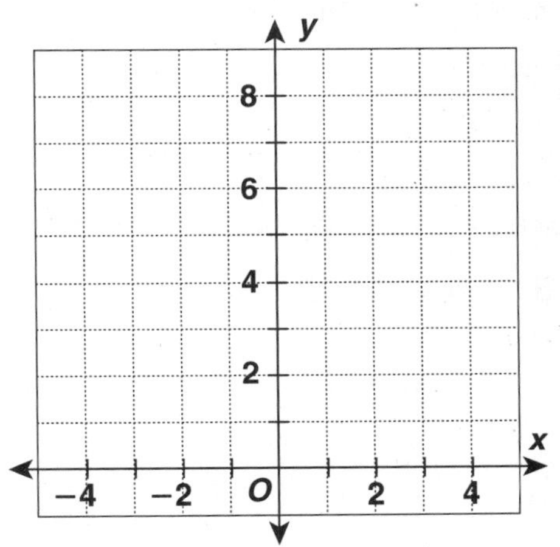

3. A forestry department introduce 500 fish to a lake. The fish are expected to increase at a rate of 35% each year. Write an exponential function to calculate the number of fish in the lake at the end of each year. Predict how many fish will be in the lake at the end of 5 years. ____________________

4. A stock valued at $756 has been declining steadily at the rate of 4% a year for the last few years. If this decline continues, predict what the value of the stock will be at the end of 3 years. ____________________

5. Todd's starting salary at his new job is $400 a week. He is promised a 3% increase in salary every year. Predict to the nearest dollar what Todd's expected yearly salary will be after working for 4 years. ____________________

Name ______________________ Date ____________ Class ____________

LESSON 13-5

Practice C

Exponential Functions

1. For each exponential function, find $f(3)$, $f(0)$, $f(-4)$.

	$f(3)$	$f(0)$	$f(-4)$
$f(x) = 4^x$			
$f(x) = 0.8^x$			
$f(x) = 15^x$			
$f(x) = 75 \bullet \left(\frac{1}{4}\right)^x$			

Write the equation of the exponential function that passes through the given points. Use the form $f(x) = p \bullet a^x$.

2. (0, 2) and (1, 8) ____________

3. (0, 5) and (1, 10) ____________

4. (0, 6) and (1, 3) ____________

Graph the exponential function of the form $f(x) = p \bullet a^x$.

5. $p = 4$, $a = 2$

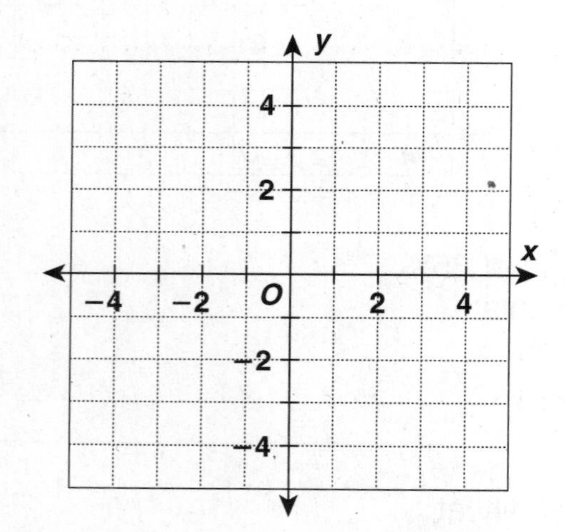

6. $p = -2$, $a = \frac{1}{2}$

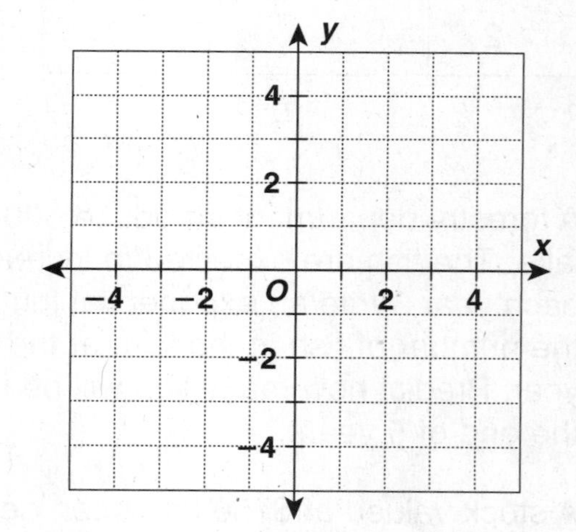

7. What percent of decrease is modeled in the exponential function $f(x) = 100 \bullet 0.92^x$? ____________

8. Mr. Harry has a rabbit farm with 10 rabbits. If the number of rabbits triples each half-year, how many rabbits will be on the farm after 3 years? ____________

Name ______________________ Date __________ Class __________

LESSON 13-5

Reteach

Exponential Functions

A function that has the input value x in the exponent is called an **exponential function.** The base number a is positive.

$$f(x) = a^x$$

Situation 1 $a > 1$, say $a = 3$

$f(x) = 3^x$

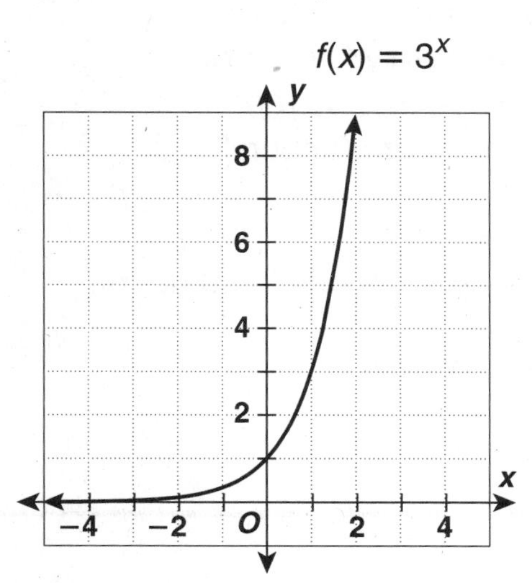

This graph rises from left to right.

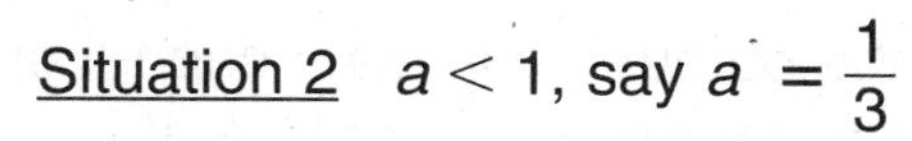

Situation 2 $a < 1$, say $a = \frac{1}{3}$

$f(x) = \left(\frac{1}{3}\right)^x$

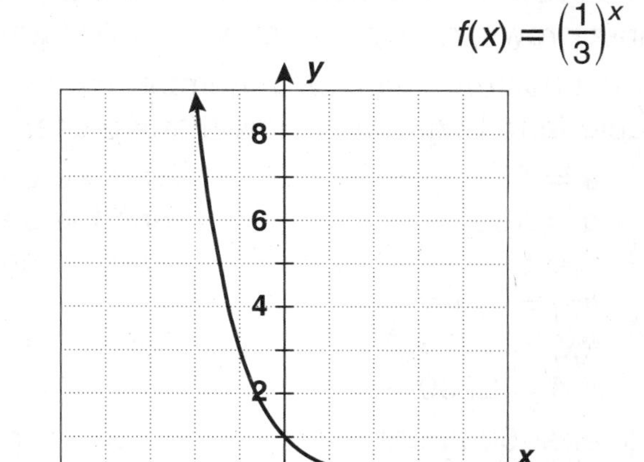

This graph falls from left to right.

For both situations: The domain is the set of all real numbers.
The range is the set of positive real numbers.
The y-intercept is 1.

Complete the table for each function.
Graph both functions on the same grid. Label each function.

1. $f(x) = 2^x$

x	y
−2	$2^{-2} = \frac{1}{2^2}$ =
−1	$2^{-1} = \frac{1}{2^1}$ =
0	$2^0 = \frac{1}{2^0}$ =
1	$2^1 =$
2	$2^2 =$

2. $f(x) = \left(\frac{1}{2}\right)^x$

x	y
−2	$\left(\frac{1}{2}\right)^{-2} = \left(\frac{2}{1}\right)^2$
−1	$\left(\frac{1}{2}\right)^{-1} = \left(\frac{2}{1}\right)$ =
0	$\left(\frac{1}{2}\right)^0 =$
1	$\left(\frac{1}{2}\right)^1 =$
2	$\left(\frac{1}{2}\right)^2 =$

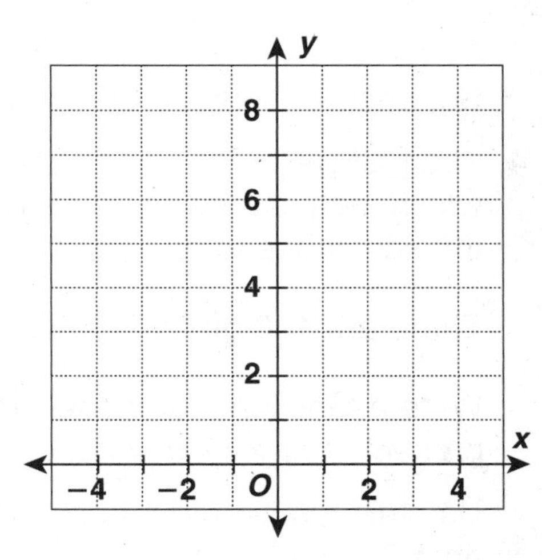

Name ________________ Date ________ Class ________

LESSON 13-5

Reteach

Exponential Functions (continued)

When an exponential function has a **starting number** p, its rule can be written $f(x) = p \cdot a^x$.

If $a > 1$, the output grows as the input grows, and the function is called an **exponential growth function**. If $a < 1$, the output shrinks as the input grows, and the function is called an **exponential decay function**.

Say a bacteria doubles in number every 10 hours, and there are 100 bacteria to begin with. How many bacteria will there be after 50 hours?

$a = 2$	Find the common ratio.
$p = 100$	Find the starting number.
$x = 5$	Find the number of 10-hour periods in 50 hours.
$f(x) = p \cdot a^x$	
$f(x) = 100 \cdot 2^5$	Substitute into the function rule.
$f(x) = 3200$	

So, after 50 hours, there will be 3200 bacteria.

Write and apply an exponential growth function.

3. Consider a bacteria that triples in number every 24 hours. $a =$ ________

   Suppose there are 40 bacteria to begin. $p =$ ________

   Write a function that models the number of bacteria present after x 24-hour periods. $f(x) =$ ________

   Use this function to predict the number of bacteria present after 48 hours. $x =$ ________

   $f(x) =$ ________________

   So, after 48 hours, there will be ______ bacteria present.

4. The half-life of Barium-131 is 12 days, which means it takes 12 days for half of the substance to decompose. $a =$ ________

   There are 100 grams to begin. $p =$ ________

   Write a function that models the number of grams present after x 12-day periods. $f(x) =$ ________

   Use this function to predict the number of grams present after 36 days. $x =$ ________

   $f(x) =$ ________ $=$ ________ $=$ ________

   So, after 36 days, there will be ______ grams of Barium-131 remaining.

Name ______________________ Date ____________ Class ____________

LESSON **13-5**

Challenge

Exponent Search

Now, you will see how to solve some exponential equations.

Solve for x: $2^{x+1} = 16$ an exponential equation

$2^{x+1} = 2^4$ Write both sides as powers of the same base.

$x + 1 = 4$ Equate the exponents.

$-1 \quad -1$ Solve the linear equation for x.

$x = 3$

Check: $2^{x+1} = 16$ Use the original equation.

$2^{3+1} \stackrel{?}{=} 16$ Substitute $x = 3$.

$2^4 \stackrel{?}{=} 16$

$16 = 16$ ✓

Solve and check each exponential equation.

1. $2^{x+3} = 64$

____

____

____

$x =$ ______

Check:

2. $3^{x+2} = 9^x$

____

____

____

$x =$ ______

Check:

3. $4^{2x+1} = 8^{2x}$

____

____

____

$x =$ ______

Check:

4. $27^x = 9^{2x-1}$

$x =$ ______

Check:

5. $\left(\frac{1}{2}\right)^x = 4$

$x =$ ______

Check:

6. $\left(\frac{1}{3}\right)^{x+1} = 27$

$x =$ ______

Check:

Name ______________________ Date ____________ Class ____________

LESSON 13-5

Problem Solving

Exponential Functions

From 1950 to 2000, the world's population grew exponentially. The function that models the growth is $f(x) = 1.056 \cdot 1.018^x$ where x is the year ($x = 50$ represents 1950) and $f(x)$ is the population in billions. Round each number to the nearest hundredth.

1. Estimate the world's population in 1950.

____

2. Estimate the world's population in 2005.

____

3. Predict the world's population in 2025.

____

4. Predict the world's population in 2050.

____

Insulin is used to treat people with diabetes. The table below shows the percent of an insulin dose left in the body at different times after injection.

Time elapsed (min)	**Percent remaining**
0	100
48	50
96	25
144	12.5

5. Which ordered pair does not represent a half-life of insulin?

   A (24, 70.71) **C** (48, 50)
   B (50, 50) **D** (72, 35.35)

6. Write an exponential function that describes the percent of insulin in the body after x half-lives.

   F $f(x) = 100\left(\frac{1}{2}\right)^x$ **H** $f(x) = 2(100)^x$
   G $f(x) = 10\left(\frac{1}{2}\right)^x$ **J** $f(x) = 48\left(\frac{1}{2}\right)^x$

7. What percent of insulin would be left in the body after 6 hours?

   A 0.25% **C** 0.55%
   B 0.39% **D** 1.56%

8. What percent of insulin would be left in the body after 9 hours?

   F 0.04% **H** 0.17%
   G 0.12% **J** 0.26%

9. A new form of insulin that is being developed has a half-life of 9 hours. Write an exponential function that describes the percent of insulin in the body after x half-lives.

   A $f(x) = 100\left(\frac{1}{2}\right)^x$ **C** $f(x) = 2(100)^x$
   B $f(x) = 9\left(\frac{1}{2}\right)^x$ **D** $f(x) = 100(9)^x$

10. What percent of the new form of insulin would be left in the body after 9 hours?

   F 12.5% **H** 50%
   G 25% **J** 75%

Name ______________________________ Date ____________ Class ____________

LESSON 13-5

Reading Strategies

Use a Context

Will started a savings plan. He saved $2 the first month. His goal was to save twice as much money each month.

Month	1	2	3	4
Amount Saved	$2	$4	$8	$16

Use the table to answer each question.

1. How much money did Will save the second month? ____________

2. How many times as great was the second month's savings as the first month's savings? ____________

3. How much money did Will save in the third month? ____________

4. How many times as great was the third month's savings as the first month's savings? ____________

Will noticed a pattern in his savings and made another table.

Month	Pattern	Savings
x	2^x	y
1	2^1	$2
2	2^2	$4
3	2^3	$8
4	2^4	$16

The function $f(x) = 2^x$ is an example of an exponential function.

Answer each question.

5. What does x stand for in this exponential function? ____________

6. What does $f(x)$ stand for in this exponential function?

____

Name ______________________________ Date ____________ Class ____________

LESSON **13-5**

Puzzles, Twisters and Teasers

Collar I.D.?

Complete the chart which shows a part of the U.S. population that is growing exponentially. Each answer has a corresponding letter. Use the corresponding letters to solve the riddle.

Americans over 100 Years Old (thousands)										
Year		2010		2030		2050		2070	2080	2090
Population	70		280		1120		4480			
	R	**G**	**E**	**O**	**V**	**L**	**I**	**D**	**C**	**N**

What do you get when you cross a dog and a phone?

A ______ ______ ______ ______ ______ ______
140 560 2240 8960 2020 35840

______ E ______ E ______ ______ E R
2000 17920 2060 2040

Name ______________________ Date __________ Class __________

LESSON 13-6

Practice A

Quadratic Functions

Complete the table for each quadratic function.

1. $f(x) = x^2 - 1$

x	$f(x) = x^2 - 1$
−3	$f(-3) = (-3)^2 - 1 = 8$
−2	
−1	
0	
1	
2	
3	

2. $f(x) = x^2 - 2x + 3$

x	$f(x) = x^2 - 2x + 3$
−3	
−2	
−1	
0	
1	
2	
3	

3. Complete the table and graph the quadratic function, $f(x) = x^2 + 4x + 2$.

x	$f(x) = x^2 + 4x + 2$
−5	
−4	
−3	
−2	
−1	
0	
1	

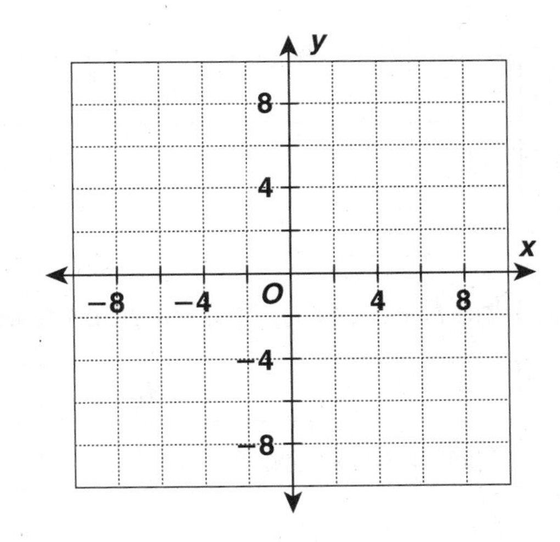

4. One number is 6 greater than another number. Their product is given by the function $f(x) = x^2 + 6x$. Which pair of numbers results in the least product?

____

Name ______________________ Date ____________ Class ____________

LESSON 13-6

Practice B

Quadratic Functions

Create a table for each quadratic function, and use it to make a graph.

1. $f(x) = x^2 - 5$

x	$f(x) = x^2 - 5$
−3	$f(-3) = (-3)^2 - 5 = 4$
−1	
0	
2	
3	

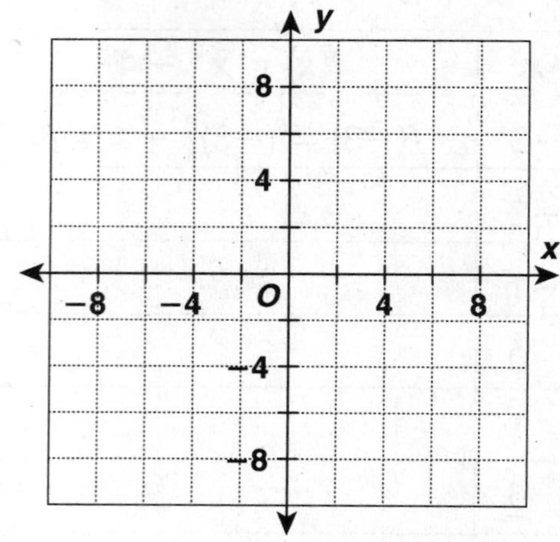

2. $f(x) = x^2 - 2x + 3$

x	$f(x) = x^2 - 2x + 3$
3	
2	
1	
0	
−1	

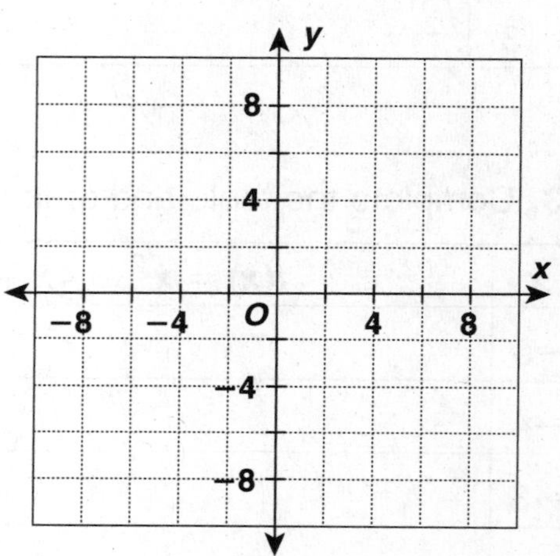

3. Find $f(-3)$, $f(0)$, $f(3)$ for each quadratic function.

	$f(-3)$	$f(0)$	$f(3)$
$f(x) = x^2 - 2x + 1$			
$f(x) = x^2 - 6$			
$f(x) = x^2 - x + 3$			

4. The function $f(t) = -4.9t^2$ gives the distance in meters that an object will fall toward Earth in t seconds. Find the distance an object will fall in 1, 2, 3, 4, and 5 seconds. (Note that the distance traveled by a falling object is shown by a negative number.)

____

Name ______________________ Date __________ Class __________

LESSON 13-6

Practice C

Quadratic Functions

Create a table for each quadratic function, and use it to make a graph.

1. $f(x) = x^2 - 5x + 6$

x	$f(x) = x^2 - 5x + 6$
−1	
0	
1	
2	
3	
4	
5	

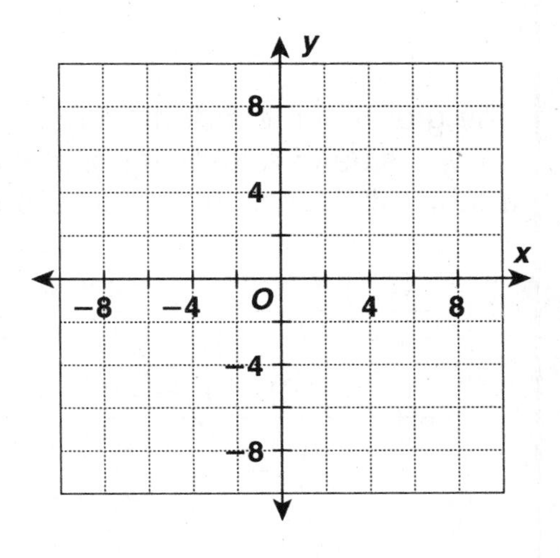

2. $f(x) = (x - 2)(x - 4)$

x	$f(x) = (x - 2)(x - 4)$
0	
1	
2	
3	
4	
5	

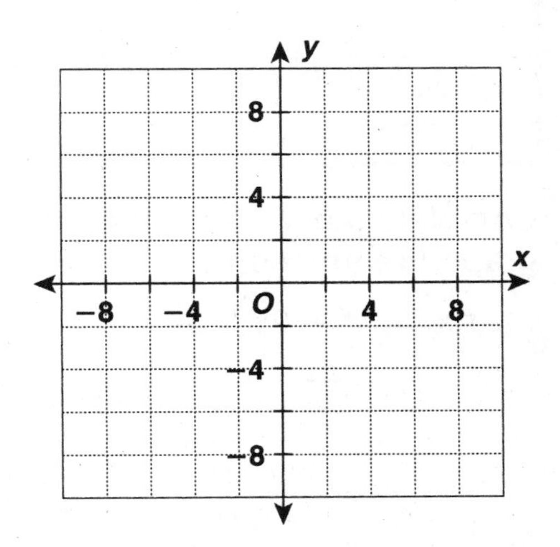

Find $f(-5)$, $f(0)$, $f(5)$ for each quadratic function.

3. $f(x) = x^2 - x - 9$

3. $f(x) = (x - 9)(x + 8)$

______________________ ______________________

4. The height in meters of a ball thrown at a certain speed is given by the function $f(t) = -4.9t^2 + 19.6t$, where t is the elapsed time in seconds. Find the height of the ball after 1, 2, 3, and 4 seconds.

____

Name ______________________ Date __________ Class __________

LESSON 13-6

Reteach

Quadratic Functions

A **quadratic function** has a variable that is squared.

general quadratic function $f(x) = ax^2 + bx + c$

square term (↑ ax^2) y-intercept (↑ c)

$f(x) = x^2 - 4x + 3$

The graph of a quadratic function is a **parabola,** a curve that falls on one side of a turning point and rises on the other. You can make a table of a function's values and use them to graph the function.

x	$f(x) = x^2 - 4x + 3$
−1	$f(-1) = (-1)^2 - 4(-1) + 3 = 8$
0	$f(0) = 0^2 - 4(0) + 3 = 3$
1	$f(1) = 1^2 - 4(1) + 3 = 0$
2	$f(2) = 2^2 - 4(2) + 3 = -1$
3	$f(3) = 3^2 - 4(3) + 3 = 0$
4	$f(4) = 4^2 - 4(4) + 3 = 3$
5	$f(5) = 5^2 - 4(5) + 3 = 8$

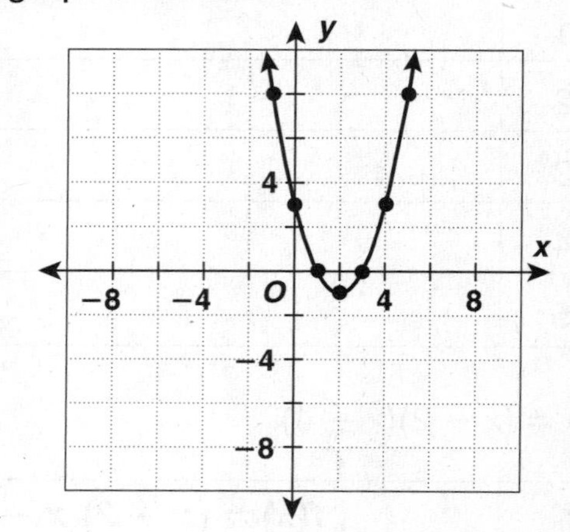

Complete the table for the quadratic function and use it to graph the function.

1. $f(x) = x^2 - 2x - 3$

x	$f(x) = x^2 - 2x - 3$
−2	$f(-2) = (\quad)^2 - 2(-2) - 3 =$
−1	$f(-1) = (\quad)^2 - 2(-1) - 3 =$
0	$f(0) = (\quad)^2 - 2(0) - 3 =$
1	$f(1) = (\quad)^2 - 2(1) - 3 =$
2	$f(2) = (\quad)^2 - 2(2) - 3 =$
3	$f(3) = (\quad)^2 - 2(3) - 3 =$
4	$f(4) = (\quad)^2 - 2(4) - 3 =$

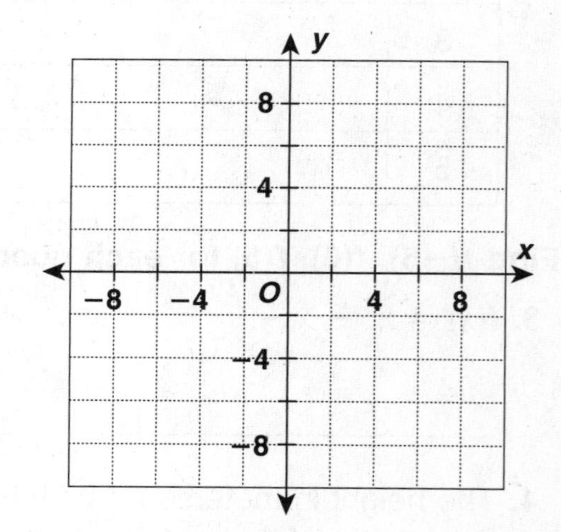

Name ______________________ Date __________ Class __________

LESSON 13-6

Reteach

Quadratic Functions (continued)

When a quadratic function is written as the product of two differences, you can read the two x-intercepts.

$f(x) = (x - r)(x - s)$
↑ r and ↑ s: x-intercepts

For the quadratic function $f(x) = (x - 3)(x + 4)$, the x-intercepts are 3 and -4.

Identify the *x*-intercepts for each function.

2. $f(x) = (x - 4)(x - 7)$

4 and ______

3. $f(x) = (x + 1)(x - 5)$

-1 and ______

4. $f(x) = (x + 2)(x + 4)$

______ and ______

Complete the table for the quadratic function and use it to graph the function. Identify the *x*-intercepts and the *y*-intercept.

5. $f(x) = (x + 1)(x - 3)$

x	$f(x) = (x + 1)(x - 3)$
-2	$f(-2) = (-2 + 1)(-2 - 3) = 5$
-1	$f(-1) = (\quad + 1)(\quad - 3) =$
0	$f(0) = (\quad + 1)(\quad - 3) =$
1	$f(1) = (\quad + 1)(\quad - 3) =$
2	$f(2) = (\quad + 1)(\quad - 3) =$
3	$f(3) = (\quad + 1)(\quad - 3) =$
4	$f(4) = (\quad + 1)(\quad - 3) =$

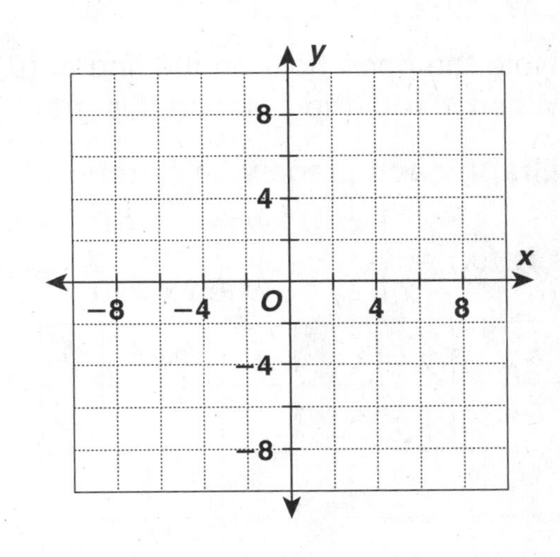

x-intercepts = ______ and ______

y-intercept = ______

Name ______________________ Date ____________ Class ____________

LESSON 13-6

Challenge

A Piece of This, a Piece of That

A function defined differently over various parts of its domain is called a **piecewise function.**

$$f(x) = \begin{cases} x - 2 & \text{when } x < 0 \\ x^2 & \text{when } x \geq 0 \end{cases}$$

This function consists of the line $y = x - 2$ when x is negative and the parabola $y = x^2$ when x is nonnegative.

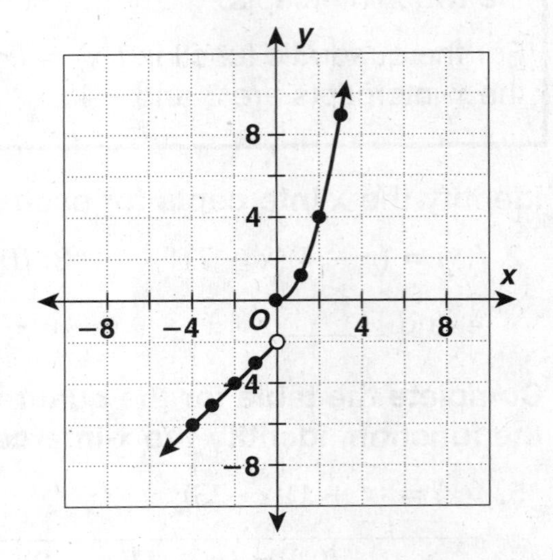

x	$x - 2$	y		x	x^2	y
−4	−4 − 2	−6		0	0^2	0
−3	−3 − 2	−5		1	1^2	1
−2	−2 − 2	−4		2	2^2	4
−1	−1 − 2	−3		3	3^2	9

Note the open hole on the line at (0, −2).
When $x = 0$, the point on this graph is on the parabola, not on the line.

Graph each piecewise function.

1. $f(x) = \begin{cases} x & \text{when } x < 0 \\ 2x^2 & \text{when } x \geq 0 \end{cases}$

x	$y = x$		x	$2x^2$	y

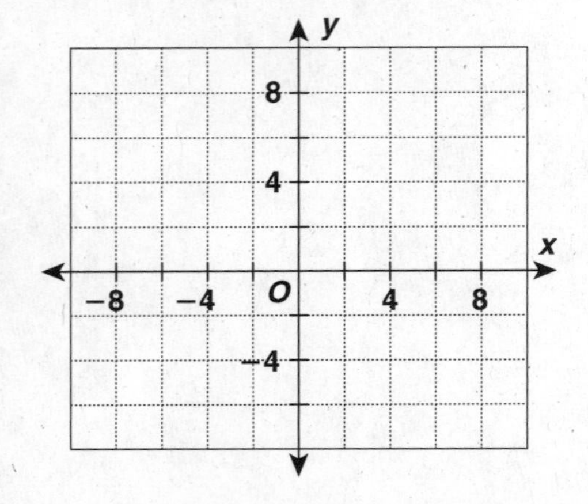

2. $f(x) = \begin{cases} x + 2 & \text{when } x < 0 \\ -x^2 & \text{when } x \geq 0 \end{cases}$

x	$x + 2$	y		x	$-x^2$	y

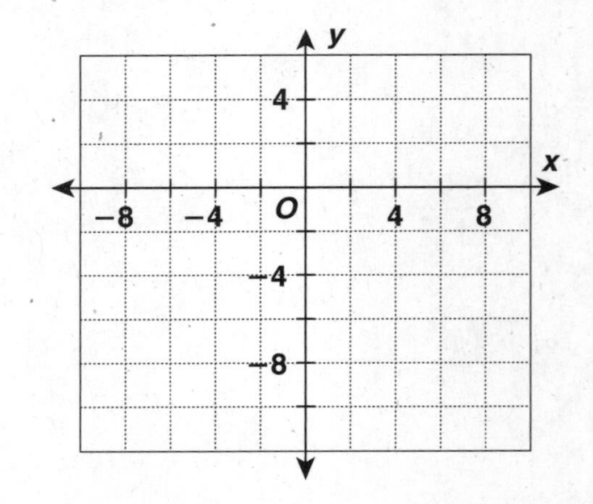

Name ______________________ Date __________ Class __________

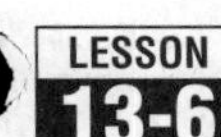

LESSON 13-6

Problem Solving

Quadratic Functions

To find the time it takes an object to fall, you can use the equation $h = -16t^2 - vt + s$ where *h* is the height in feet, *t* is the time in seconds, *v* is the initial velocity, and *s* is the starting height in feet. Write the correct answer.

1. If a construction worker drops a tool from 240 feet above the ground, how many feet above the ground will it be in 2 seconds? Hint: $v = 0$, $s = 240$.

____

2. How long will it take the tool in Exercise 1 to hit the ground? Round to the nearest hundredth.

____

3. The Gateway Arch in St. Louis, Missouri is the tallest manmade memorial. The arch rises to a height of 630 feet. If you throw a rock down from the top of the arch with a velocity of 20 ft/s, how many feet above the ground will the rock be in 2 seconds?

____

4. Will the rock in exercise 3 hit the ground within 6 seconds of throwing it?

____

The average monthly rainfall for Seattle, Washington can be approximated by the equation $f(x) = 0.147x^2 - 1.890x + 7.139$ where *x* is the month (January: $x = 1$, February, $x = 2$, etc.) and $f(x)$ is the monthly rainfall in inches. Choose the letter for the best answer.

5. What is the average monthly rainfall in Seattle for the month of January?

   A 3.7 in   **C** 7.6 in

   B 5.4 in   **D** 9.2 in

6. What is the average monthly rainfall in Seattle for the month of April?

   F 0.2 in   **H** 1.9 in

   G 1.4 in   **J** 2.8 in

7. What is the average monthly rainfall in Seattle for the month of August?

   A 1.1 in   **C** 5.6 in

   B 1.4 in   **D** 6.8 in

8. In what month does it rain the least in Seattle, Washington?

   F May   **H** July

   G June   **J** August

Name ______________________ Date ____________ Class ____________

LESSON 13-6

Reading Strategies

Analyze Information

A **quadratic function** contains a variable to the second power.

This equation is a quadratic function. → $y = x^2 + 1$

A table of values and the graph for this quadratic function are shown below.

x	$x^2 + 1$	y	(x, y)
−2	$(-2)^2 + 1$	5	(−2, 5)
−1	$(-1)^2 + 1$	2	(−1, 2)
0	$0^2 + 1$	1	(0, 1)
1	$1^2 + 1$	2	(1, 2)
2	$2^2 + 1$	5	(2, 5)

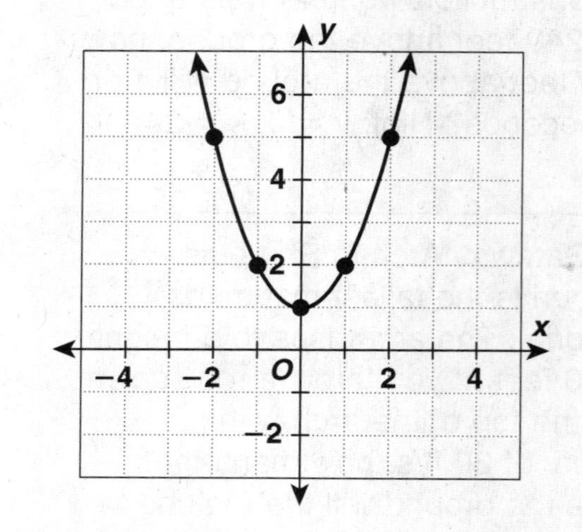

The shape of this graph is called a **parabola.** The graphs of quadratic functions are always parabolas.

Note how pairs of points that make up the parabola are located directly opposite each other on the graph.

Answer each question.

1. What is the power of one variable in a quadratic function?

____

2. What name is given to the shape of the graph for a quadratic function?

____

3. In your own words, how would you describe the shape of the graph of a quadratic function?

____

4. The coordinates (−1, 2) identify a point on the left side of the parabola. Write the coordinates that identify the opposite point on the graph.

____

Name ______________________________ Date ____________ Class ____________

Puzzles, Twisters and Teasers

Paws for the Cause!

Solve each equation. Then plot the points from the table on the graph and connect them with a smooth curve. Each point has a corresponding letter. Use the letters to solve the riddle.

x	$f(x) = x^2 - 2$	y
−3	$(-3)^2 - 2$	
−2	$(-2)^2 - 2$	
−1	$(-1)^2 - 2$	
0	$(0)^2 - 2$	
1	$(1)^2 - 2$	
2	$(2)^2 - 2$	
3	$(3)^2 - 2$	

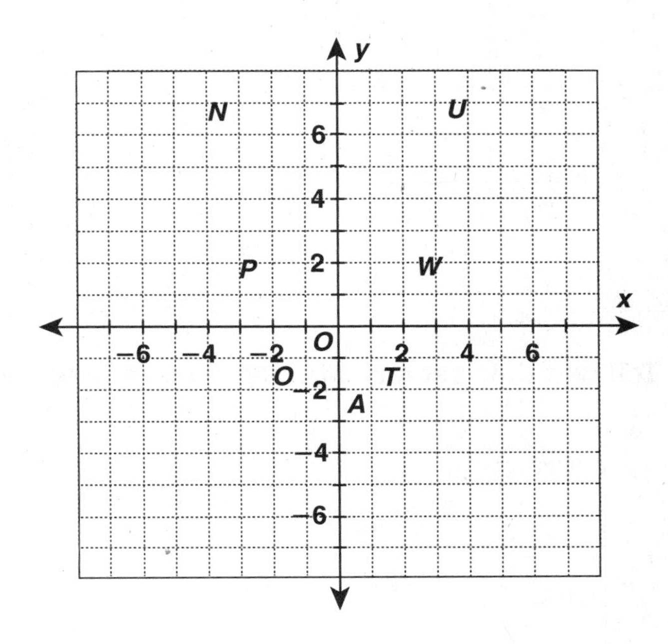

How does a dog stop a VCR?

By pressing the ______ ______ ______ S

(−2, 2) (0, −2) (2, 2)

B ______ ______ T ______ ______

(3, 7) (1, −1) (−1, −1) (−3, 7)

Name ______________________ Date ____________ Class ____________

LESSON 13-7

Practice A

Inverse Variation

Complete the table for each inverse variation.

1. $xy = 60$

x	y
2	
3	
4	
5	

2. $xy = 1$

x	y
−8	
−2	
$\frac{1}{4}$	
3	

Tell whether each relationship is an inverse variation.

3.

x	y
6	4
3	8
2.5	9.6
−12	−2

____

____

4.

x	y
9	8
3	24
0.25	288
0.06	120

____

____

5. Complete the table and graph the inverse variation function.

x	y
2	6
3	
−8	
−4	
−2	

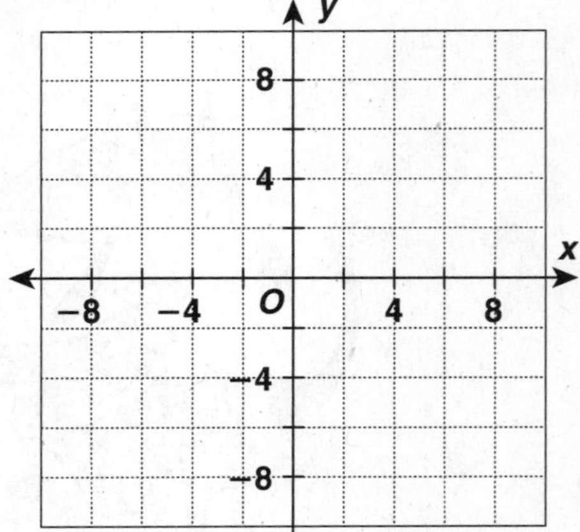

6. If x varies inversely as y and $x = 6$ when $y = 7$, what is the constant of variation?

____

Name ______________________ Date __________ Class __________

LESSON 13-7

Practice B

Inverse Variation

Tell whether each relationship is an inverse variation.

1. The table shows the length and width of certain rectangles.

Length	6	8	12	16	24
Width	8	6	4	3	2

____

2. The table shows the number of days needed to paint a house for the size of the work crew.

Crew Size	2	3	4	5	6
Days of Painting	21	14	10.5	8.5	7

____

3. The table shows the time spent traveling at different speeds.

Hours	5	6	8	9	12
mi/h	72	60	45	40	30

____

Graph each inverse variation function.

4. $f(x) = \frac{4}{x}$

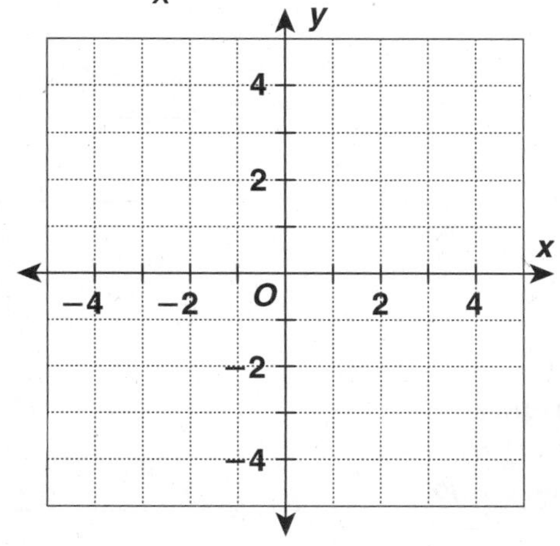

5. $f(x) = \frac{5}{x}$

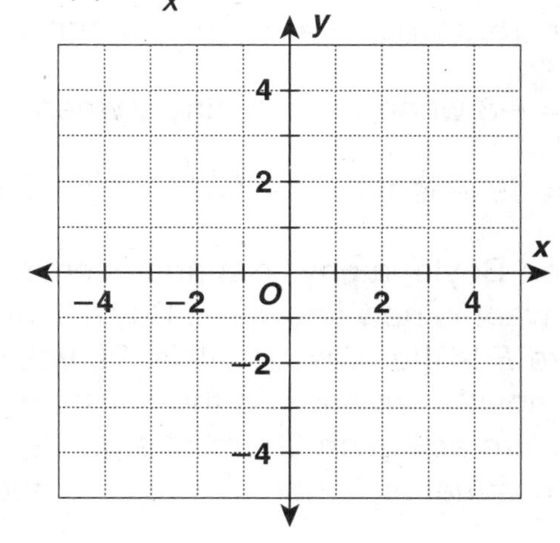

6. Amperes (abbreviated amp) measure the strength of electric current. An ohm is the unit of electrical resistance. In an electric circuit, the current varies inversely as the resistance. If the current is 24 amps when the resistance is 20 ohms, find the inverse variation function and use it to find the resistance in ohms when the current is 40 amps. ______________

Name ______________________ Date __________ Class __________

LESSON 13-7

Practice C

Inverse Variation

Find the inverse variation function, given that x and y vary inversely.

1. y is 4 when x is 22 ____________
2. y is 16 when x is 4 ____________
3. y is -12 when x is 18 ____________
4. y is -15 when x is -50 ____________
5. y is 0.5 when x is 37 ____________
6. y is $\frac{1}{3}$ when x is 2115 ____________
7. If y varies inversely with x and if $y = -2$ when $x = -12$, find the constant of variation. ____________
8. If y varies inversely with x and if $y = -\frac{2}{3}$ when $x = 48$, find the constant of variation. ____________

For exercises 9–14, assume that y varies inversely as x.

9. If $y = -6$ when $x = -2$, find y when $x = 5$. ____________
10. If $y = 200$ when $x = -\frac{1}{2}$, find y when $x = -2.5$. ____________
11. If $y = 72.6$ when $x = 15$, find y when $x = 33$. ____________
12. If $y = 15.5$ when $x = 3$, find y when $x = 5$. ____________
13. If $y = -8$ when $x = 3.4$, find y when $x = 4$. ____________
14. If $y = 49$ when $x = 14$, find y when $x = -7$. ____________
15. Robert Boyle, a physicist and chemist is credited with what is now known as Boyle's law, $PV = k$, where P is the pressure, V is the volume measured in atmospheres and k is the constant of proportionality. Pressure acting on 30 m^3 of a gas is reduced from 2 to 1 atmospheres. What new volume does the gas occupy? ____________
16. Suppose pressure acting on 30 m^3 of a gas is increased from 1 to 2 atmospheres. What new volume does the gas occupy? ____________

Name ______________________________ Date ____________ Class ____________

LESSON 13-7

Reteach

Inverse Variation

Two quantities **vary inversely** if their product is constant.

y varies inversely as *x*. $xy = k$ ← constant of variation

$y = \frac{k}{x}$ ← function of inverse variation

To determine if two data sets vary inversely, check for a constant product.

x	1	2	4	8	−1	−2	−4	−8
y	8	4	2	1	−8	−4	−2	−1

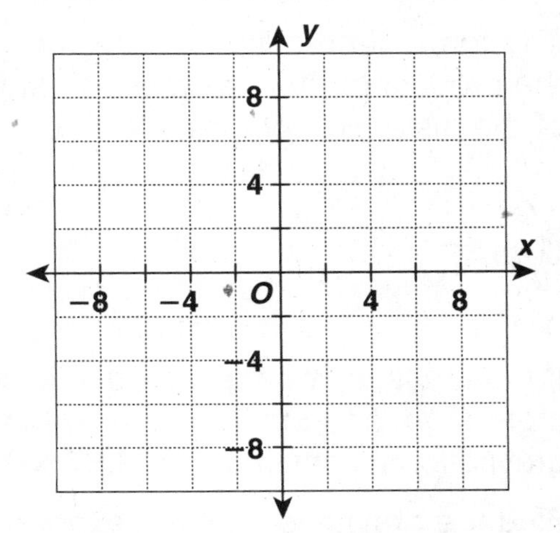

$xy = 1(8) = 2(4) = 4(2) = 8(1) = -1(-8)$
$= -2(-4) = -4(-2) = -8(-1) = 8$

So, *y* varies inversely as *x*.

The constant of variation is 8.

The function of inverse variation is $y = \frac{8}{x}$.

The graph of an inverse variation is two curves and is not defined for $x = 0$.

Tell whether the data shows inverse variation. If there is a constant product, identify it and write the function of inverse variation.

1.

x	5	4	2	1	−1	−2
y	20	16	8	4	−4	−8

constant product? ________________

If yes, function. ________________

2.

x	1	2	3	−3	−2
y	12	6	4	−4	−6

constant product? ________________

If yes, function. ________________

Write the function of inverse variation. Then graph it.

3.

x	1	2	3	6
y	6	3	2	1
x	−1	−2	−3	−6
y	−6	−3	−2	−1

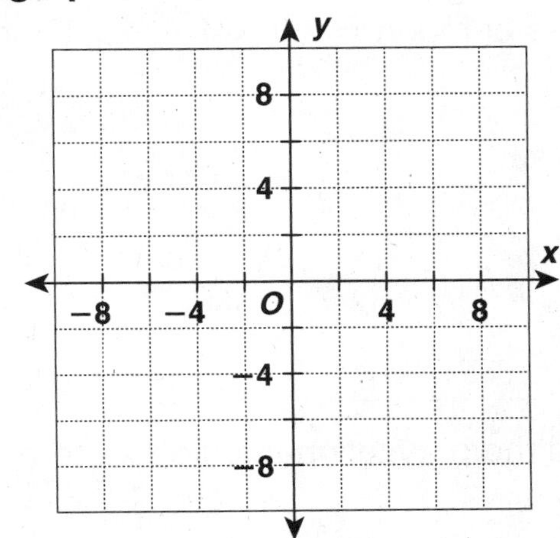

function: ________________

Name ______________________ Date ____________ Class ____________

LESSON 13-7

Challenge

When an Apple Fell on His Head!

The English physicist Sir Isaac Newton is said to have recognized the force of gravity as a result of having an apple fall from a tree under which he was seated. He later reasoned that any two objects attract one another gravitationally.

Newton's Law of Gravitation

The force of attraction F between two particles varies directly with the product of their masses, m_1 and m_2, and inversely as the square of the distance r between their centers.

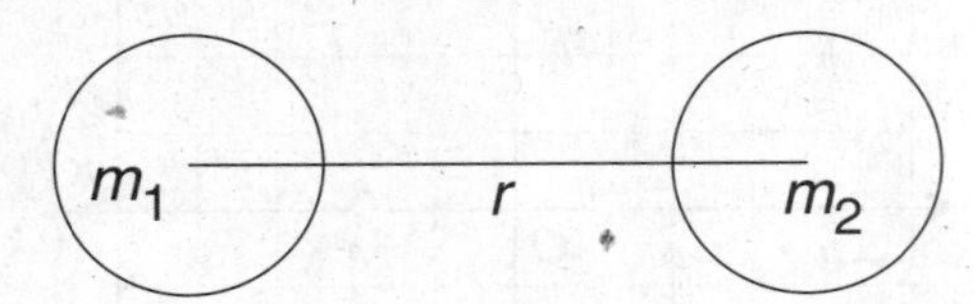

$F = \frac{G m_1 m_2}{r^2}$ G is the universal gravitational constant.

If the masses are expressed in kilograms, kg, the distance in meters, m, and the force in newtons, N, the value of the universal gravitational constant $G = 6.67 \times 10^{-11}$ N • m^2 / kg^2.

Find the gravitational force exerted by the Earth on a 1-kg mass on its surface.

mass of object, $m_1 = 1.0$ kg
mass of Earth, $m_2 = 6.0 \times 10^{24}$ kg
distance between = radius of Earth = 6.4×10^6 m

$$F = \frac{G m_1 m_2}{r^2} = \frac{6.67 \times 10^{-11}(1.0)(6.0 \times 10^{24})}{(6.4 \times 10^6)^2}$$

$$= \frac{(6.67)(1.0)(6.0) \times 10^{-11+24}}{(6.4)^2 \times 10^{6 \cdot 2}}$$

$$= \frac{40.02 \times 10^{13}}{40.96 \times 10^{12}} = 0.977 \times 10 = 9.8 \text{ N}$$

1. Find the gravitational force exerted by the moon on the Earth.
mass of moon = 7.3×10^{22} kg, Earth-moon distance = 3.8×10^7 m

$F =$ ____________

2. Find the gravitational force exerted by the Earth on the moon.

$F =$ ______________________________

Name ______________________ Date __________ Class __________

LESSON 13-7

Problem Solving

Inverse Variation

For a given focal length of a camera, the f-stop varies inversely with the diameter of the lens. The table below gives the f-stop and diameter data for a focal length of 400 mm. Round to the nearest hundredth.

f-stop	diameter (mm)
1	400
2	200
4	100
8	50
16	25
32	12.5

1. Use the table to write an inverse variation function.

____

2. What is the diameter of a lens with an f-stop of 1.4?

____

3. What is the diameter of a lens with an f-stop of 11?

____

4. What is the diameter of a lens with an f-stop of 22?

____

The inverse square law of radiation says that the intensity of illumination varies inversely with the square of the distance to the light source.

5. Using the inverse square law of radiation, if you halve the distance between yourself and a fire, by how much will you increase the heat you feel?

   A 2
   B 4
   C 8
   D 16

6. Using the inverse square law of radiation, if you double the distance between a radio and the transmitter, how will it affect the signal intensity?

   F $\frac{1}{4}$ as strong
   G $\frac{1}{2}$ as strong
   H twice as strong
   J 4 times stronger

7. Using the inverse square law of radiation, if you increase the distance between yourself and a light by 4 times, how will it affect the light's intensity?

   A $\frac{1}{16}$ as strong
   B $\frac{1}{4}$ as strong
   C $\frac{1}{2}$ as strong
   D twice as strong

8. Using the inverse square law of radiation if you move 3 times closer to a fire, how much more intense will the fire feel?

   F $\frac{1}{3}$ as strong
   G 3 times stronger
   H 9 times stronger
   J 27 times stronger

Name ______________________ Date ____________ Class ____________

LESSON **13-7**

Reading Strategies

Use a Context

Jason drew different sketches for a dog pen with an area of 24 square units. Notice that as the length of the pen increases, the width of the pen decreases.

Length	Width	Area
2	12	24
3	8	24
4	6	24

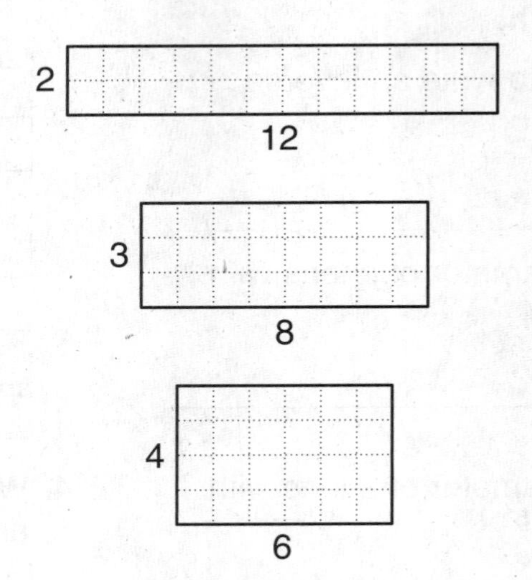

As one variable increases (length), the other variable decreases (width). The product of the variables (length and width) remains the same. (The area of each pen is always 24 square units.) This relationship is called an **inverse variation.**

Use the information above to answer each question.

1. Which dimension of the pen increases?

____

2. Which dimension of the pen decreases?

____

3. What stayed the same when the dimensions of the pen changed?

____

4. What was the area of each pen?

____

5. What name is given to the relationship between a variable that increases and a variable that decreases with a constant product?

____

Name ______________________ Date ____________ Class ____________

LESSON 13-7

Puzzles, Twisters and Teasers

Bounce This Around!

Fill in the blanks to complete the chart. Each answer has a corresponding letter.

Use the letters to solve the riddle.

Hint: $y = \frac{180}{x}$

Barn Construction Days and Size of Work Crew										
Crew Size			5		15		30	36	45	
Number	90	60		18		9				2
of Days	**E**	**T**	**N**	**R**	**I**	**A**	**L**	**M**	**O**	**P**

What lived in prehistoric times, had very sharp teeth, and went boing-boing?

A T-Rex on a ___ ___ ___ ___ ___ ___ ___ ___ ___ ___

3 10 20 5 90 4 6 12 36 2

LESSON 13-1
Practice A
Terms of Arithmetic Sequences

Find the common difference for each arithmetic sequence.

1. 2, 4, 6, 8, 10, … **2**
2. 19, 13, 7, 1, −5, … **−6**
3. −3, −6, −9, −12, −15, … **−3**
4. −13, −9, −5, −1, 3, … **4**
5. 1.1, 2.2, 3.3, 4.4, 5.5, … **1.1**
6. 2, $\frac{3}{2}$, 1, $\frac{1}{2}$, 0, … **$-\frac{1}{2}$**

Find the next three terms in each sequence.

7. 15, 11, 7, 3, −1, … **−5, −9, −13**
8. −22, −28, −34, −40, −46, … **−52, −58, −64**
9. −18, −13, −8, −3, 2, … **7, 12, 17**
10. 41, 32, 23, 14, 5, … **−4, −13, −22**
11. −2.1, −3.7, −5.3, −6.9, −8.5, … **−10.1, −11.7, −13.3**
12. $\frac{5}{3}$, $\frac{4}{3}$, 1, $\frac{2}{3}$, $\frac{1}{3}$, … **0, $-\frac{1}{3}$, $-\frac{2}{3}$**

Find the given term in each arithmetic sequence.

13. 10th term: 15, 24, 33, 42, 51, … **96**
14. 16th term: 5, 3, 1, −1, −3, … **−25**
15. James is given 75 vocabulary words the first week in English class. He learns 10 words the first day and five more each day after that. How many days will it take James to learn all 75 vocabulary words? **14 days**

LESSON 13-1
Practice B
Terms of Arithmetic Sequences

Determine if each sequence could be arithmetic. If so, give the common difference.

1. 18, 20, 22, 24, 26, … **2**
2. 48, 42, 36, 30, 24, … **−6**
3. 15, 30, 60, 120, 240, … **no**
4. 10.4, 8.3, 6.2, 4.1, 2, … **−2.1**
5. $\frac{1}{3}$, $\frac{1}{9}$, $\frac{1}{27}$, $\frac{1}{81}$, $\frac{1}{243}$, … **no**
6. 83, 66, 49, 32, 15, … **−17**
7. 8.1, 2.7, 0.9, 0.3, 0.1, … **no**
8. $\frac{2}{3}$, $\frac{4}{3}$, 2, $\frac{8}{3}$, $\frac{10}{3}$, … **$\frac{2}{3}$**
9. −58, −35, −12, 11, 34, … **23**

Find the given term in each arithmetic sequence.

10. 14th term: 60, 68, 76, 84, 92, … **164**
11. 35th term: 3.5, 3.8, 4.1, 4.4, 4.7, … **13.7**
12. 21st term: 103, 84, 65, 46, 27, … **−277**
13. 22nd term: −2, −5, −8, −11, −14, … **−65**
14. 16th term: 73, 44, 15, −14, −43, … **−362**
15. 50th term: −9, 2, 13, 24, 35, … **530**
16. 19th term: −87, −78, −69, −60, −51, … **75**
17. 25th term: $3\frac{1}{4}$, $3\frac{1}{2}$, $3\frac{3}{4}$, 4, $4\frac{1}{4}$, … **$9\frac{1}{4}$**
18. A cook started with 26 ounces of special sauce. She used 1.4 ounces of the sauce in each of a number of dishes and had 2.2 ounces left over. How many dishes did she make with the sauce? **17 dishes**
19. Kuang started the basketball season with 54 points in his career. He scores 3 points more each game he plays. How many games will it take for him to have scored a total of 132 points in his basketball career? **26 games**

LESSON 13-1
Practice C
Terms of Arithmetic Sequences

Find the given term in each arithmetic sequence.

1. 14th term: 7, 2, −3, −8, −13, … **−58**
2. 9th term: −12, −4, 4, 12, 20, … **52**
3. 20th term: $\frac{3}{4}$, $\frac{1}{4}$, $-\frac{1}{4}$, $-\frac{3}{4}$, $-1\frac{1}{4}$, … **$-8\frac{3}{4}$**
4. 11th term: 3.5, 5.2, 6.9, 8.6, 10.3, … **20.5**

Write the next three terms of each arithmetic sequence.

5. $\frac{3}{5}$, $\frac{6}{5}$, $\frac{9}{5}$, $\frac{12}{5}$, 3, … **$\frac{18}{5}$, $\frac{21}{5}$, $\frac{24}{5}$**
6. 3, −8, −19, −30, −41, … **−52, −63, −74**
7. 8.9, 10.3, 11.7, 13.1, 14.5, … **15.9, 17.3, 18.7**

Write the first five terms of each arithmetic sequence.

8. $a_1 = 7$, $d = -5$ **7, 2, −3, −8, −13, …**
9. $a_1 = \frac{3}{4}$, $d = -\frac{1}{2}$ **$\frac{3}{4}$, $\frac{1}{4}$, $-\frac{1}{4}$, $-\frac{3}{4}$, $-\frac{5}{4}$, …**
10. $a_1 = 3.5$, $d = 1.7$ **3.5, 5.2, 6.9, 8.6, 10.3, …**
11. The 12th term of an arithmetic sequence is 100. The common difference is 8. What are the first five terms of the arithmetic sequence? **12, 20, 28, 36, 44**
12. The 25th term of an arithmetic sequence is −157. The common difference is −6. What are the first five terms of the arithmetic sequence? **−13, −19, −25, −31, −37**
13. Neleh opens the bowling season with an average of 135. Each week she raises her average 2 points. At the end of the 20 week season, what will her average be? **173**

LESSON 13-1
Reteach
Terms of Arithmetic Sequences

In an **arithmetic sequence,** the difference between terms is constant. The difference is called the **common difference.**

This is an arithmetic sequence with a common difference of 3.

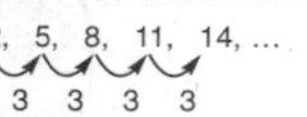

This is not an arithmetic sequence since there is no common difference.

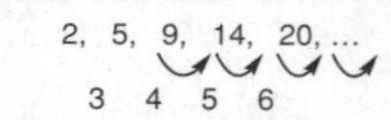

Complete to determine if each sequence is arithmetic.

1. 20, 16, 12, 8, 4, … differences: −4, **−4**, **−4**, **−4**; arithmetic? **yes**
2. 1, 2, 4, 8, 16, … differences: 1, **2**, **4**, **8**; arithmetic? **no**
3. 0.1, 0.2, 0.3, 0.4, … differences: **0.1**, **0.1**, **0.1**; arithmetic? **yes**
4. $\frac{1}{2}$, 1, $\frac{3}{2}$, 2, $\frac{5}{2}$, … differences: **$\frac{1}{2}$**, **$\frac{1}{2}$**, **$\frac{1}{2}$**, **$\frac{1}{2}$**; arithmetic? **yes**
5. 2, $\frac{3}{2}$, 1, $\frac{1}{2}$, 0, … differences: **$-\frac{1}{2}$**, **$-\frac{1}{2}$**, **$-\frac{1}{2}$**, **$-\frac{1}{2}$**; arithmetic? **yes**
6. 3, 1, 0, $-\frac{1}{2}$, $-\frac{1}{4}$, … differences: **−2**, **−1**, **$-\frac{1}{2}$**, **$\frac{1}{4}$**; arithmetic? **no**

You can use the common difference to find any term in an arithmetic sequence.

4, 6, 8, 10, 12, … This arithmetic sequence has a common difference of 2.
(differences: 2 2 2 2)

This is the 1st term of the sequence.	4
For the 2nd term, add the common difference × 1	4 + 2 × 1 = 6
For the 3rd term, add the common difference × 2	4 + 2 × 2 = 8
For the 4th term, add the common difference × 3	4 + 2 × 3 = 10
For the 5th term, add the common difference × 4	4 + 2 × 4 = 12
For the *n*th term, add the common difference × (*n* − 1)	4 + 2 × (*n* − 1)

Complete to find the given term of the arithmetic sequence 4, 6, 8, 10, 12, ….

7. the 9th term: 4 + 2 × **8** = **20**
8. the 20th term: 4 + 2 × **19** = **42**
9. the 100th term: 4 + 2 × **99** = **202**

LESSON 13-1 Reteach
Terms of Arithmetic Sequences (continued)

You can use a formula to find the nth term, a_n, of an arithmetic sequence with common difference d. $a_n = a_1 + (n - 1)d$

Find the 20th term of this arithmetic sequence: 2, 5, 8, 11, 14, ...

$a_n = a_1 + (n - 1)d$
$a_{20} = 2 + (20 - 1)3$
$a_{20} = 2 + (19)3 = 2 + 57 = 59$

Complete to find the given term of each arithmetic sequence.

10. 28, 34, 40, 46, 52, ...
Find the 18th term.
$n =$ 18, $a_1 =$ 28, $d =$ 6
$a_{18} =$ 28 + (18 − 1) 6
= 28 + 102 = 130

11. $\frac{1}{8}, \frac{1}{4}, \frac{3}{8}, \frac{1}{2}, \frac{5}{8}, \ldots$
Find the 25th term.
$n =$ 25, $a_1 = \frac{1}{8}$, $d = \frac{1}{8}$
$a_{25} = \frac{1}{8} + (25 - 1)\frac{1}{8}$
$= \frac{1}{8} + 3 = 3\frac{1}{8}$

You can use the same formula to find other missing information.

What term of the arithmetic sequence 0.25, 0.50, 0.75, ... is 6.5?

Assign values to the variables.	$a_1 = 0.25$, $d = 0.25$, $a_n = 6.5$, $n = ?$
Substitute into the formula.	$a_n = a_1 + (n - 1)d$
Solve for n.	$6.5 = 0.25 + (n - 1)0.25$
Multiply.	$6.5 = 0.25 + 0.25n - 0.25$
Combine like terms.	$6.5 = 0.25n$
Divide.	$\frac{6.5}{0.25} = \frac{0.25n}{0.25}$
So, 6.5 is the 26th term.	$26 = n$

Complete to answer the question.

12. What term of the arithmetic sequence 8, 16, 24, 32, ... is 112?

$a_n = a_1 + (n - 1)d$
112 = 8 + $(n - 1)$ 8 Substitute.
$a_1 =$ 8
112 = 8 + 8n − 8 Solve for n.
$d =$ 8
112 = 8n
$a_n =$ 112
$\frac{112}{8} = \frac{8n}{8}$
So, 112 is the 14th term
14 = n

LESSON 13-1 Challenge
Learn from "The Prince of Mathematics"

When the German mathematician Karl Gauss was a schoolboy, his teacher gave the problem of summing the integers from 1 through 100, hoping it would keep the class quiet. But young Karl wrote the correct answer after only a few seconds. How did he do it?

This is the sum. $S = 1 + 2 + 3 + \ldots + 98 + 99 + 100$
Reverse the numbers. $S = 100 + 99 + 98 + \ldots + 3 + 2 + 1$
Add vertically. $2S = 101 + 101 + 101 + \ldots + 101 + 101 + 101$

This sum contains one hundred addends of 101.

$2S = 100(101)$

$\frac{2S}{2} = \frac{100(101)}{2}$

$S = \frac{100}{2}(101) = 50(101) = 5050$

Gauss had come upon a method for finding the sum of any number of terms in an arithmetic sequence. $S_n = \frac{n}{2}(a_1 + a_n)$

Applying the formula to the original problem:

Substitute $n = 100$, $a_1 = 1$, $a_n = 100$ $S_{100} = \frac{100}{2}(1 + 100) = 50(101) = 5050$

1. Now that you know how to find the sum of the first 100 integers, and you know what the sum is, can you just divide by 2 to find the sum of the first 50 even integers? of the first 50 odd integers? Explain.

No; 5050 ÷ 2 = 2525
sum of first 50 even: $S_{50} = \frac{50}{2}(2 + 100) = 25(102) = 2550$
sum of first 50 odd: $S_{50} = \frac{50}{2}(1 + 99) = 25(100) = 2500$

2. Find the sum of the first 750 integers.

$S_{750} = \frac{750}{2}(1 + 750) = 375(751) = 281,625$

3. Find the sum of the first 100 terms of this arithmetic sequence: 3, 6, 9, 12, ... (Hint: First find the 100th term.)

$a_{100} = 3 + (100 - 1)3 = 300$
$S_{100} = \frac{100}{2}(3 + 300) = 50(303) = 15,150$

LESSON 13-1 Problem Solving
Terms of Arithmetic Sequences

A section of seats in an auditorium has 18 seats in the first row. Each row has two more seats than the previous row. There are 25 rows in the section. Write the correct answer.

1. List the number of seats in the second, third and fourth rows of the section.
20, 22, 24

2. How many seats are in the 10th row?
36

3. How many seats are in the 15th row?
46

4. In which row are there 32 seats?
8th row

For 5–10, refer to the table below, which shows the boiling temperature of water at different altitudes. Choose the letter of the correct answer.

Altitude (thousands of feet)	Boiling point of water (°F)
1	210.2
2	208.4
3	206.6
4	204.8
5	203

5. What is the common difference?
(A) −1.8°F C −2.8°F
B 1.8°F D 6°F

6. According to the table, what would be the boiling point of water at an altitude of 10,000 feet?
F 192.2°F H 226.4°F
(G) 194°F J 228.2°F

7. According to the table, what would be the boiling point of water at an altitude of 15,000 feet?
A 181.4°F (C) 185°F
B 183.2°F D 235.4°F

8. Estimate the boiling point of water in Jacksonville, Florida, which has an elevation of 0 feet.
F 0°F (H) 212°F
G 208.4°F J 213.8°F

9. The highest point in the United States is Mt. McKinley, Alaska, with an elevation of 20,320 feet. Estimate the boiling point of water at the top of Mt. McKinley.
A 172.4°F C 244.4°F
(B) 176°F D 246.2°F

10. At which elevation will the boiling point of water be less than 150°F?
F 28,000 ft H 32,000 ft
G 30,000 ft (J) 35,000 ft

LESSON 13-1 Reading Strategies
Focus On Vocabulary

A **sequence** is a list of numbers arranged in a certain order. Each number in a sequence is called a **term.**

An **arithmetic sequence** is a sequence in which the same number is added to each term to get the next term. The number that is added is called the **common difference.**

1st Term	2nd Term	3rd Term	4th Term
4	7	10	13

(differences: 3, 3, 3)

Answer the following questions.

1. What do you call the list of numbers above? a sequence
2. What do you call each number in the list? a term
3. How can you tell whether a sequence is an arithmetic sequence?
The difference between one term and the next is always the same.
4. Is the list of numbers above an arithmetic sequence? Explain.
yes, because the difference between terms is the same
5. What is the common difference in the arithmetic sequence above? 3
6. How do you find the common difference?
Find the number that is added to a term to get the next term, or subtract a term from the next term.
7. If the arithmetic sequence above continues, what number will come next in the sequence? 16

LESSON **13-1**

Puzzles, Twisters and Teasers

Take It or Leave It!

Find and circle words from the list in the word search (horizontally, vertically or diagonally). Find a word that answers the riddle. Circle it and write it on the line.

sequence term arithmetic common order
difference finite pattern subscript position

```
G V H B D I F F E R E N C E
S E Q U E N C E D S T Z X A
U J N H Y U O A Q W E C V R
B K J H G F M E R T R B N I
S B G T R W M Y U I M M L T
I U G O L V O R D E R F K H
P A T T E R N O P L K I J M
F O O T P R I N T S Q N H E
C D S U B S C R I P T I G T
N H Y T G B V F R E C T F I
P O S I T I O N A S D E D C
```

The more you take, the more you leave behind. What are they?

FOOTPRINTS

LESSON **13-2**

Practice A

Terms of Geometric Sequences

Find the common ratio for each of the following geometric sequences.

1. 5, 10, 20, 40, 80, ... **2**
2. 3, 15, 75, 375, 1875, ... **5**
3. 1, 6, 36, 216, 1296, ... **6**
4. 0.5, 1.5, 4.5, 13.5, 40.5, ... **3**
5. 810, 270, 90, 30, 10, ... $\frac{1}{3}$
6. 72, 7.2, 0.72, 0.072, 0.0072, ... **0.1**

Find the given term in each geometric sequence.

7. 7th term: 2, 6, 18, 54, ... **1,458**
8. 10th term: 25, 5, 1, 0.2, ... **0.0000128**
9. 8th term: 8, 4, 2, 1, ... $\frac{1}{16}$
10. 6th term: 3, 4.5, 6.75, 10.125, ... **22.78125**

Find the next three terms of each geometric sequence.

11. 2, 10, 50, 250, 1250, ... **6250; 31,250; 156,250**
12. 4, 24, 144, 864, 5184, ... **31,104; 186,624; 1,119,744**
13. 375, 75, 15, 3, 0.6, ... **0.12, 0.024, 0.0048**
14. $\frac{1}{3125}, \frac{1}{625}, \frac{1}{125}, \frac{1}{25}, \frac{1}{5}, \ldots$ **1, 5, 25**
15. 1.8, 3.6, 7.2, 14.4, 28.8, ... **57.6, 115.2, 230.4**
16. 6804, 2268, 756, 252, 84, ... **28, $9.\overline{33}$, $3.\overline{11}$**
17. Julie is doing an experiment. She is studying a cell that triples in number every hour. She started the experiment with 24 cells. How many cells are there at the end of 4 hours?

   1944 cells

LESSON **13-2**

Practice B

Terms of Geometric Sequences

Determine if each sequence could be geometric. If so, give the common ratio.

1. 4, 16, 64, 256, 1024, ... **4**
2. 3, $\frac{3}{2}, \frac{3}{4}, \frac{3}{8}, \frac{3}{16}, \ldots$ $\frac{1}{2}$
3. 5, 10, 15, 20, 25, ... **not**
4. 3, 18, 108, 648, 3888, ... **6**
5. 1250, 125, 12.5, 1.25, 0.125, ... **0.1**
6. 10, 15, 22.5, 33.75, 50.625, ... **1.5**
7. 36, 12, 4, $\frac{4}{3}, \frac{4}{9}, \ldots$ $\frac{1}{3}$
8. 1440, 720, 240, 60, 12, ... **not**
9. 9, 3, 1, 0.5, 0.25, ... **not**

Find the given term in each geometric sequence.

10. 6th term: 25, 75, 225, 675, ... **6075**
11. 10th term: 320, 160, 80, 40, ... **0.625**
12. 9th term: 4.5, 9, 18, 36, ... **1152**
13. 7th term: 0.02, 0.2, 2, 20, ... **20,000**
14. 12th term: $\frac{1}{1000}, \frac{1}{100}, \frac{1}{10}, 1, \ldots$ **100,000,000**
15. 8th term: $\frac{3}{8}, \frac{3}{4}, \frac{3}{2}, 3, \ldots$ **48**
16. In an experiment a population of flies triples every week. The experiment starts with 12 flies. How many flies will there be by the end of week 5?

    2916 flies
17. A small business earned $21 in its first month. It quadrupled this amount each month for the next several months. How much did the business earn in the 4th month?

    $1344

LESSON **13-2**

Practice C

Terms of Geometric Sequences

Find the given term in each geometric sequence.

1. 8th term: $\frac{2}{81}, \frac{4}{27}, \frac{8}{9}, 5\frac{1}{3}, \ldots$ **6912**
2. 7th term: 0.004, 0.04, 0.4, 4, ... **4,000**

Find the next three terms of each geometric sequence.

3. $a_1 = 3$, common ratio = 7 **21, 147, 1029**
4. $a_1 = 800$, common ratio = 0.4 **320, 128, 51.2**
5. $a_1 = \frac{3}{8}$, common ratio = 2 $\frac{3}{4}, \frac{3}{2}, 3$
6. $a_1 = 7.6$, common ratio = 8 **60.8, 486.4, 3891.2**

Find the first five terms of each geometric sequence.

7. $a_1 = 250$, $r = 0.6$ **250, 150, 90, 54, 32.4**
8. $a_1 = 0.16$, $r = 5$ **0.16, 0.8, 4, 20, 100**
9. $a_1 = \frac{2}{81}$, $r = 6$ $\frac{2}{81}, \frac{4}{27}, \frac{8}{9}, 5\frac{1}{3}, 32$
10. $a_1 = 0.004$, $r = 10$ **0.004, 0.04, 0.4, 4, 40**
11. $a_1 = 12$, $r = 9$ **12, 108, 972, 8748, 78,732**
12. $a_1 = 320$, $r = \frac{1}{4}$ **320, 80, 20, 5, $1\frac{1}{4}$**
13. Find the 1st term of a geometric sequence with 5th term 11.664 and common ratio 0.6. **90**
14. Find the 1st and 7th terms of a geometric sequence with 3rd term $\frac{4}{9}$ and 4th term $\frac{8}{27}$.

    1st term = 1, 7th term = $\frac{64}{729}$
15. Find the 1st term of a geometric sequence with 10th term −1024 and common ratio −2. **2**
16. Water is leaking from a water tower. On the first day two gallons were lost. The leak is getting progressively worse and the amount of water lost triples each day. How many gallons would be lost on the 8th day? **4374 gallons**

LESSON 13-2

Reteach

Terms of Geometric Sequences

In a **geometric sequence,** the ratio of one term to the next is constant. The ratio is called **common ratio**.

This is a geometric sequence with a common ratio of $\frac{3}{1} = 3$.

2, 6, 18, 54, 162, ...

$\frac{3}{1}$ $\frac{3}{1}$ $\frac{3}{1}$ $\frac{3}{1}$

This is not a geometric sequence since there is no common ratio.

2, 4, 2, 4, 2, ...

$\frac{2}{1}$ $\frac{1}{2}$ $\frac{2}{1}$ $\frac{1}{2}$

Complete to determine if each sequence is geometric. Write *yes* or *no*.

1. $\frac{1}{125}, \frac{1}{25}, \frac{1}{5}, 1, 5, \ldots$

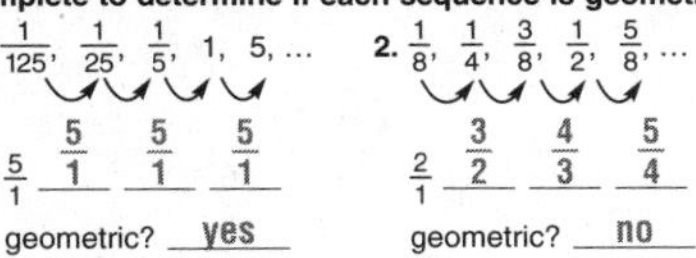

$\frac{5}{1}$ $\frac{5}{1}$ $\frac{5}{1}$ $\frac{5}{1}$

geometric? yes

2. $\frac{1}{8}, \frac{1}{4}, \frac{3}{8}, \frac{1}{2}, \frac{5}{8}, \ldots$

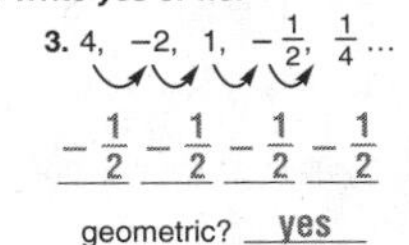

$\frac{2}{1}$ $\frac{3}{2}$ $\frac{4}{3}$ $\frac{5}{4}$

geometric? no

3. $4, -2, 1, -\frac{1}{2}, \frac{1}{4}, \ldots$

$-\frac{1}{2}$ $-\frac{1}{2}$ $-\frac{1}{2}$ $-\frac{1}{2}$

geometric? yes

You can use the common ratio to find any term in a geometric sequence.

2, 8, 32, 128, ...This geometric sequence has a common ratio of $\frac{4}{1}$ or 4.

4 4 4 4

This is the 1st term of the sequence.	2
Multiply it by the common ratio to get the 2nd term.	$2 \times 4 = 8$
Multiply it by the square of the common ratio to get the 3rd term.	$2 \times 4^2 = 32$
Multiply it by the cube of the common ratio to get the 4th term.	$2 \times 4^3 = 128$
To get the *n*th term, multiply the 1st term by the common ratio raised to the $(n - 1)$ power.	$2 \times 4^{n-1}$

Complete to find the given term of the geometric sequence 2, 8, 32, 128, ...

4. 5th term $2 \times 4^{4} = 512$

5. 7th term $2 \times 4^{6} = 8192$

6. 6th term $2 \times 4^{5} = 2048$

7. 10th term $2 \times 4^{9} = 524{,}288$

8. 8th term $2 \times 4^{7} = 32{,}768$

9. 9th term $2 \times 4^{8} = 131{,}072$

LESSON 13-2

Reteach

Terms of Geometric Sequences (continued)

You can use a formula to find the *n*th term, a_n, of a geometric sequence with common ratio *r*. $a_n = a_1 \cdot r^{n-1}$

Find the 7th term of this geometric sequence: 3, 6, 12, 24, 48, ...

$a_n = a_1 \cdot r^{n-1}$ Find *r*. $r = \frac{6}{3} = 2$

$a_7 = 3 \cdot 2^{7-1}$ Substitute $n = 7$, $a_1 = 3$, $r = 2$.

$a_7 = 3 \cdot 2^{7-1} = 3 \cdot 2^6 = 3 \cdot 64 = 192$ The 7th term is 192.

Find the given term of each geometric sequence.

10. 1, 10, 100, 1000, . . .
Find the 9th term.

$n = 9$, $a_1 = 1$, $r = \frac{10}{1} = 10$

$a_9 = 1 \cdot 10^{9-1} = 1 \cdot 10^8$

$a_9 = 1 \cdot 100{,}000{,}000$

$= 100{,}000{,}000$

11. 1.1, 1.21, 1.331, 1.4641, . . .
Find the 7th term.

$n = 7$, $a_1 = 1.1$, $r = \frac{1.21}{1.1} = 1.1$

$a_7 = 1.1 \times 1.1^{7-1} = 1.1 \times 1.1^6$

$a_7 = 1.1 \times 1.771561$

$= 1.9487171$

Which sequence has the greater 20th term? by how much?

1000, 1050, 1100, 1150, ...
arithmetic sequence, $d = 50$

$a_{20} = 1000 + (20 - 1)50$

$a_{20} = 1000 + (19)50 = 1000 + 950$

$a_{20} = 1950$

2, 4, 8, 16, ...
geometric sequence, $r = 2$

$a_{20} = 2 \cdot 2^{20-1}$

$a_{20} = 2 \cdot 2^{19} = 2 \cdot 524{,}288 = 1{,}048{,}576$

$a_{20} = 1{,}048{,}576$

The 20th term of this geometric sequence is greater by 1,046,626.

Determine whether each sequence is arithmetic or geometric and find its 15th term.

12. $2, 1, \frac{1}{2}, \frac{1}{4}, \ldots$

sequence is geometric, $r = \frac{1}{2}$

$a_n = a_1 \cdot r^{n-1}$

$a_{15} = 2 \cdot \frac{1}{2}^{15-1} = 2 \cdot \frac{1}{16{,}384}$

$a_{15} = \frac{1}{8192}$

13. $\frac{5}{2}, 3, \frac{7}{2}$

sequence is arithmetic, $d = \frac{1}{2}$

$a_n = a_1 + (n - 1)d$

$a_{15} = \frac{5}{2} + (15 - 1)\frac{1}{2}$

$= \frac{5}{2} + (14)\frac{1}{2} = \frac{5}{2} + 7$ $a_{15} = 9\frac{1}{2}$

LESSON 13-2

Challenge

What's That Sum?

You can use a formula to find the sum of *n* terms of a geometric sequence with common ratio *r*. $S_n = \frac{a_1 - a_1 r^n}{1 - r}$

Find the sum of the first 5 terms of the geometric sequence 5, 15, 45, ...

$S_n = \frac{a_1 - a_1 r^n}{1 - r}$ Find *r*. $r = \frac{15}{5} = 3$

$S_5 = \frac{5 - 5 \cdot 3^5}{1 - 3}$ Substitute $n = 5$, $a_1 = 5$, $r = 3$.

$S_5 = \frac{5 - 5 \cdot 243}{1 - 3} = \frac{5 - 1215}{1 - 3} = \frac{-1210}{-2} = 605$

So, the sum of the first 5 terms of the sequence is 605.

Check: 5 + 15 + 45 + 135 + 405 = 605

Use the formula to find each sum. Check your work by adding the terms with a calculator.

1. 32, 16, 8, ...
Find the sum of the first 6 terms.

$S_n = \frac{a_1 - a_1 r^n}{1 - r}$

$S_6 = \frac{32 - 32\left(\frac{1}{2}\right)^6}{1 - \frac{1}{2}}$

$S_6 = \frac{32 - 32\left(\frac{1}{64}\right)}{1 - \frac{1}{2}}$

$S_6 = \frac{32 - \frac{1}{2}}{1 - \frac{1}{2}} = \frac{31\frac{1}{2}}{\frac{1}{2}}$

$S_6 = 63$

Check:

32 + 16 + 8 + 4 + 2 + 1 = 63

2. −3, 15, −75, ...
Find the sum of the first 5 terms.

$S_n = \frac{a_1 - a_1 r^n}{1 - r}$

$S_5 = \frac{-3 - (-3)(-5)^5}{1 - (-5)}$

$S_5 = \frac{-3 + 3(-3125)}{1 + 5}$

$S_5 = \frac{-3 - 9375}{6} = \frac{-9378}{6}$

$S_5 = -1563$

Check:

−3 + 15 + (−75) + 375 + (−1875) = −1563

LESSON 13-2

Problem Solving

Terms of Geometric Sequences

For Exercises 1–2, determine if the sequence could be geometric. If so, find the common ratio. Write the correct answer.

1. A computer that was worth \$1000 when purchased was worth \$800 after six months, \$640 after a year, \$512 after 18 months, and \$409.60 after two years.

Could be geometric; 0.8

2. A student works for a starting wage of \$6.00 per hour. She is told that she can expect a \$0.25 raise every six months.

Not geometric

3. A piece of paper that is 0.01 inches thick is folded in half repeatedly. If the paper were folded 6 times, how thick would the result be?

0.64 inches

4. A vacuum pump removes one-half of the air in a container with each stroke. How much of the original air is left in the container after 8 strokes?

$\frac{1}{256}$

For exercises 5–8, assume that the cost of a college education increases an average of 5% per year. Choose the letter of the correct answer.

5. If the in-state tuition at the University of Florida is \$2256 per year, what will the tuition be in 10 years?
A \$3174.24
B \$3333.14
C \$3499.80
(D) \$3674.79

6. If it costs \$3046 per year for tuition for a Virginia resident at the University of Virginia now, how much will tuition be in 8 years?
F \$4183.26
G \$4286.03
(H) \$4500.33
J \$4725.35

7. If it costs \$25,839 per year in tuition to attend Northwestern University now, how much will tuition be in 5 years?
A \$31,407.47
(B) \$32,977.84
C \$37,965.97
D \$42,483.72

8. If you start attending Northwestern University in 5 years and attend for 4 years, how much will you spend in total for tuition?
(F) \$142,138.61
G \$135,370.12
H \$131,911.36
J \$169,934.88

LESSON 13-2

Reading Strategies

Analyze Information

A **geometric sequence** is formed by *multiplication*. The same number is multiplied by each term in the sequence to get the next term. That number is called a **constant**.

Term	1st	2nd	3rd	4th
Value	2	6	18	54

Use the table to answer each question.

1. What is the 1st term in this geometric sequence? 2
2. What is the 2nd term in this geometric sequence? 6
3. What number do you multiply 2 by to get 6? 3
4. What is the 3rd term in this geometric sequence? 18
5. What number do you multiply 6 by to get 18? 3
6. What is the constant in this geometric sequence? 3

Multiplying by a constant in a geometric sequence results in a **common ratio** from one term to the next.

Use the table above to answer each question.

7. What is the ratio of the 2nd term to the 1st term? 3
8. What is the ratio of the 3rd term to the 2nd term? 3
9. What is the common ratio of this geometric sequence? 3

LESSON 13-2

Puzzles, Twisters and Teasers

Itching for Answers!

Find the given term in each sequence. Use the answers to solve the riddle.

1. 12th term: 3, 6, 12, 24, 48 ... 6144 **T**
2. 8th term: 1, 4, 16, 64, 256 ... 16,384 **S**
3. 9th term: −4, −2, 0, 2, 4 ... 12 **F**
4. 6th term: $\frac{1}{2}$, 1, 2, 4 ... 16 **E**
5. 7th term: 3, 6, 12, 24 ... 192 **L**
6. 5th term: 1, 1.5, 2.25, 3.375 ... 5.0625 **C**
7. 10th term: $\frac{1}{3}, \frac{-1}{3}, \frac{1}{3}, \frac{-1}{3}, \frac{1}{3}$ ... $\frac{-1}{3}$ **K**
8. 6th term: 96, 48, 24, 12, 6 ... 3 **A**
9. 10th term: 5, −5, 5, −5, 5 ... −5 **R**
10. 6th term: 1, 2, 4, 8 ... 32 **M**

Where should you never take a dog?

To a F L E A M A R K E T

F	L	E	A	M	A	R	K	E	T
12	192	16	3	32	3	−5	$-\frac{1}{3}$	16	6144

LESSON 13-3

Practice A

Other Sequences

Complete the table by finding the first and second differences in each sequence.

1.

Sequence	2	5	11	20	32	47	65
1st Differences	3	6	9	12	15	18	
2nd Differences	3	3	3	3	3		

2.

Sequence	5	12	21	32	45	60	77
1st Differences	7	9	11	13	15	17	
2nd Differences	2	2	2	2	2		

Use first and second differences to find the next three terms in each sequence.

3. 3, 8, 18, 33, 53, ... 78, 108, 143
4. 5, 6, 8, 12, 19, ... 30, 46, 68
5. 6, 8, 12, 18, 26, ... 36, 48, 62
6. 1.5, 2.5, 5.5, 10.5, 17.5, ... 26.5, 37.5, 50.5
7. 7, 9, 13, 20, 31, ... 47, 69, 98
8. $\frac{1}{2}$, 1, 2, $3\frac{1}{2}$, $5\frac{1}{2}$, ... 8, 11, $14\frac{1}{2}$

Find the first five terms of each sequence defined by the given rule.

9. $a_n = 2n + 6$ 8, 10, 12, 14, 16
10. $a_n = \frac{n-2}{n}$ $-1, 0, \frac{1}{3}, \frac{1}{2}, \frac{3}{5}$
11. $a_n = \frac{3n-1}{2}$ $1, 2\frac{1}{2}, 4, 5\frac{1}{2}, 7$

12. The rule of a sequence is to square the number of each term's position and add 2. Find the first six terms of the sequence.
3, 6, 11, 18, 27, 38

LESSON 13-3

Practice B

Other Sequences

Use first and second differences to find the next three terms in each sequence.

1. 3, 6, 10, 15, 21, ... 28, 36, 45
2. 11, 14, 18, 25, 37, ... 56, 84, 123
3. 10, 16, $22\frac{1}{3}$, 29, 36, ... $43\frac{1}{3}$, 51, 59
4. 14.5, 22.5, 31, 40, 49.5, ... 59.5, 70, 81

Give the next three terms in each sequence using the simplest rule you can find.

5. 6, 7, 10, 19, 38, ... 71, 122, 195
6. 0.5, 2, 4.5, 8, 12.5, ... 18, 24.5, 32
7. 36, 55, 80, 111, 148, ... 191, 240, 295
8. 3, 10, 21, 36, 55, ... 78, 105, 136
9. 1, 6, 15, 28, 45, ... 66, 91, 120
10. 0, 11, 30, 57, 92, ... 135, 186, 245

Find the first five terms of each sequence defined by the given rule.

11. $a_n = \frac{n^2+2}{n}$ $3, 3, 3\frac{2}{3}, 4\frac{1}{2}, 5\frac{2}{5}$
12. $a_n = \frac{5n-2}{n+1}$ $1\frac{1}{2}, 2\frac{2}{3}, 3\frac{1}{4}, 3\frac{3}{5}, 3\frac{5}{6}$
13. $a_n = \frac{3n^2}{n+2}$ $1, 3, 5\frac{2}{5}, 8, 10\frac{5}{7}$

14. Suppose *a*, *b*, and *c* are three consecutive numbers in the Fibonacci sequence. Complete the following table and guess the pattern.

a, b, c	ab	bc
1, 1, 2	1	2
2, 3, 5	6	15
5, 8, 13	40	104
13, 21, 34	273	714
34, 55, 89	1870	4895

The difference of *bc* and *ab* is the square of *b*.

LESSON 13-3

Practice C
Other Sequences

Give the next three terms in each sequence using the simplest rule you can find.

1. 14, 25, 36, 47, 58, ...
   69, 80, 91
2. 9, 12, 17, 24, 33, ...
   44, 57, 72
3. $1\frac{1}{5}$, $4\frac{4}{5}$, $10\frac{4}{5}$, $19\frac{1}{5}$, 30, ...
   $43\frac{1}{5}$, $58\frac{4}{5}$, $76\frac{4}{5}$
4. 10.5, 84, 283.5, 672, 1312.5, ...
   2268, 3601.5, 5376

Find the first five terms of each sequence defined by the given rule.

5. $a_n = 2n^2 + 6$
   8, 14, 24, 38, 56
6. $a_n = \frac{12n - 5}{n}$
   7, $9\frac{1}{2}$, $10\frac{1}{3}$, $10\frac{3}{4}$, 11
7. $a_n = \frac{n^2 - n}{2n}$
   0, $\frac{1}{2}$, 1, $1\frac{1}{2}$, 2
8. $a_n = \frac{4n^2 + 2n}{n + 2}$
   2, 5, $8\frac{2}{5}$, 12, $15\frac{5}{7}$
9. $a_n = \frac{3n^2 - 2n}{2n + 1}$
   $\frac{1}{3}$, $1\frac{3}{5}$, 3, $4\frac{4}{9}$, $5\frac{10}{11}$
10. $a_n = \frac{6n^2 - 2n + 3}{n + 2}$
   $2\frac{1}{3}$, $5\frac{3}{4}$, $10\frac{1}{5}$, $15\frac{1}{6}$, $20\frac{3}{7}$

11. Suppose *a, b, c, d, e, f, g, h, i,* and *j* are ten consecutive numbers in the Fibonacci sequence. Complete the following table and guess the pattern.

a, b, c, d, e, f, g, h, i, j	*11 • g*	*a + b + c + d + e + f + g + h + i + j*
1, 1, 2, 3,5, 8, 13, 21, 34, 55	143	143
55, 89, 144, 233, 377, 610, 987, 1597, 2584, 4181	10,857	10,857

The product 11 • *g* is equal to the sum *a + b + c + d + e + f + g + h + i + j.*

12. The sum of any 10 consecutive terms in the Fibonacci sequence is divisible by what number?
   11

LESSON 13-3

Reteach
Other Sequences

Differences can help you find patterns in some sequences.

Find the next number in the sequence: 1, 6, 15, 28, 45, ...

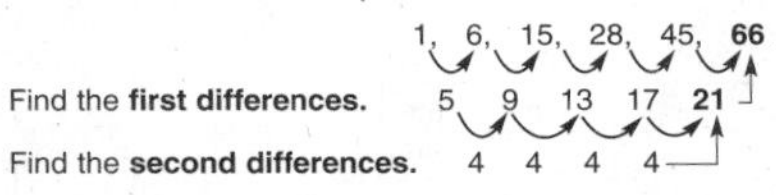

Find the **first differences.**

Find the **second differences.**

Complete to find the next term in each sequence.

1. 1, 4, 9, 16, 25, ...
   3 5 7 9
   2 2 2
   The next term in the sequence is:
   25 + 11 = 36
2. 1, 8, 21, 40, 65, ...
   7 13 19 25
   6 6 6
   The next term in the sequence is:
   65 + 31 = 96

A rule is used to define a sequence.

Write a rule for this sequence: $\frac{1}{2}, \frac{2}{3}, \frac{3}{4}, \frac{4}{5}, \frac{5}{6}, \ldots$

A possible rule is that the numerator of a term is the number of that terms position, and the denominator is 1 more than the numerator.

This can be written algebraically as $a_n = \frac{n}{n+1}$.

Using this rule, the 10th term of the sequence is $a_{10} = \frac{10}{10+1} = \frac{10}{11}$.

Use the given rule to write the 5th and 10th terms.

3. 1, 8, 27, 64, ...
   $a_n = n^3$
   $a_5 = (\ 5\ )^3 = 125$
   $a_{10} = (\ 10\ )^3 = 1000$
4. 1, 3, 6, 10, ...
   $a_n = \frac{n(n+1)}{2}$
   $a_5 = \frac{(5)(5+1)}{2} = 15$
   $a_{10} = \frac{(10)(10+1)}{2} = 55$
5. 1, −3, 1, −3, ...
   $a_n = 2(-1)^{n+1} - 1$
   $a_5 = 2(-1)^{5+1} - 1$
   $= 1$
   $a_{10} = 2(-1)^{10+1} - 1$
   $= -3$

LESSON 13-3

Challenge
Follow the Short-Cut

The Greek capital letter *sigma*, ∑, is used to mean "take the sum of what follows."

What follows ∑ is a general term that is a rule for each term of a *summation.*

The general term is written with an *index*, shown by one letter (often, *n*).

The *limits* for the index are written above ∑ (upper limit) and below ∑ (lower limit).

The first term of the summation is formed by substituting the lower limit for the index into the general term.

Each succeeding term of the summation is formed by using successive integral values of the index, until the upper limit is reached.

$\sum_{n=1}^{5} 2^n$ This notation means take the sum of terms of the form 2^n for consecutive integral values of *n* beginning with *n* = 1 and ending with *n* = 5.

$$\sum_{n=1}^{5} 2^n = 2^1 + 2^2 + 2^3 + 2^4 + 2^5$$

$$\sum_{n=1}^{5} 2^n = 2 + 4 + 8 + 16 + 32 = 62$$

Evaluate each summation.

1. $\sum_{n=1}^{5} 4n$
   $= 4(1) + 4(2) + 4(3) + 4(4) + 4(5)$
   $= 4 + 8 + 12 + 16 + 20$
   $= 60$
2. $\sum_{n=1}^{3} (n^2 + 1)$
   $= (1^2 + 1) + (2^2 + 1) + (3^2 + 1)$
   $= (2) + (5) + (10)$
   $= 17$
3. $\sum_{n=1}^{4} \frac{6}{n}$
   $= \frac{6}{1} + \frac{6}{2} + \frac{6}{3} + \frac{6}{4}$
   $= 6 + 3 + 2 + 1.5$
   $= 12.5$
4. $\sum_{n=1}^{4} \frac{n}{n+1}$
   $= \frac{1}{1+1} + \frac{2}{2+1} + \frac{3}{3+1} + \frac{4}{4+1}$
   $= \frac{1}{2} + \frac{2}{3} + \frac{3}{4} + \frac{4}{5}$
   $= \frac{30}{60} + \frac{40}{60} + \frac{45}{60} + \frac{48}{60} = 2\frac{43}{60}$

LESSON 13-3

Problem Solving
Other Sequences

A toy rocket is launched and the height of the rocket during its first four seconds is recorded. Write the correct answer.

Time (sec)	Height (ft)
0	0
1	176
2	320
3	432
4	512
5	560
6	576
7	560

1. Find the first differences for the rocket's heights.
   176, 144, 112, 80
2. Find the second differences.
   −32, −32, −32
3. Use the first and second differences to predict the height of the rocket at 5, 6, and 7 seconds.
4. What is the maximum height of the rocket?
   576 ft
5. When will the rocket hit the ground?
   12 seconds after takeoff

For exercises 6–9, refer to the table below, which shows the number of diagonals for different polygons. Choose the letter for the correct answer.

Polygon	Sides	Diagonals
Triangle	3	0
Quadrilateral	4	2
Pentagon	5	5
Hexagon	6	9
Heptagon	7	14

6. What are the first differences for the diagonals?
   A 1, 1, 1, 1
   B 3, 2, 0, 3, 7
   (C) 2, 3, 4, 5
   D 2, 7, 14, 23
7. What are the second differences?
   (F) 1, 1, 1
   G 1, 2, 3, 4
   H 5, 7, 8
   J 7, 9, 11, 13
8. How many diagonals does a nonagon (9 sides) have?
   A 21
   B 24
   (C) 27
   D 32
9. Which rule will give the number of diagonals *d* for *s* sides?
   F $d = \frac{s(s+1)}{2}$
   G $d = (s - 3)(s - 2) - 1$
   (H) $d = \frac{s(s-3)}{2}$
   J $d = (s - 3)(s - 2)$

LESSON 13-3
Reading Strategies
Look for a Pattern

To continue a sequence or find the rule for the sequence, **look for a pattern.** You can analyze increases (or decreases) from one term to the next to find the pattern.

Term	1st	2nd	3rd	4th	5th
Value	2	5	9	14	20

3 4

Use the table to answer each question.

1. What is the 1st term in the sequence? **2**
2. What is the 2nd term of the sequence? **5**
3. What is the increase in value from the 1st term to the 2nd term? **3**
4. What is the increase in value between the 2nd and 3rd terms? **4**
5. What is the increase between the 3rd and 4th terms? **5**
6. List in order the increases you found between one number and the next in the sequence above. **3,4,5**
7. What pattern do you see in the increases?

   Possible answer: Each term increases by one more than the previous term.
8. What will the 6th term in the sequence be? **27**
9. What will the 10th term in the sequence be? **65**

LESSON 13-3
Puzzles, Twisters and Teasers
Puzzle Pattern!

Across

1. In a ________ sequence, add the two previous terms to find the next term.
3. If you do not see a pattern in the first diffrences, try finding the ________ differences.
5. To continue a sequence, look for a ________.
7. Some sequences are defined by a given ________.
8. First and second differences can help you find patterns in some ________.
9. When looking for a sequence with no given rule, try the ________ rule first.

Down

2. Sometimes a(n) ________ rule is used to define a sequence.
4. To begin, look for a pattern using first ________.
6. In a sequence with no given rule, you cannot ________ what the next term will be.

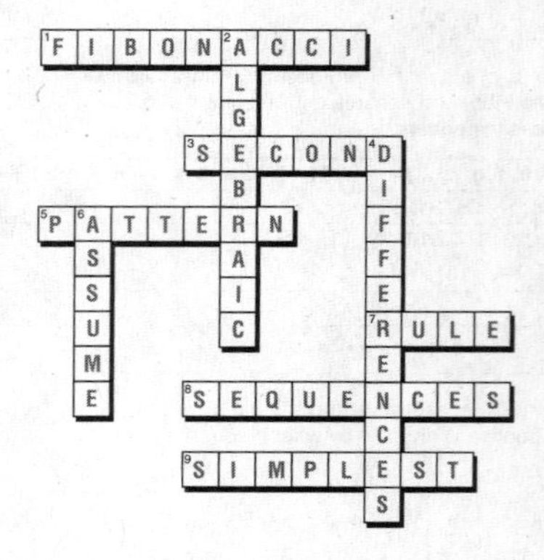

LESSON 13-4
Practice A
Linear Functions

Graph each linear and write a rule for the function.

1.

x	y
−2	0
−1	−1
0	−2
1	−3
2	−4

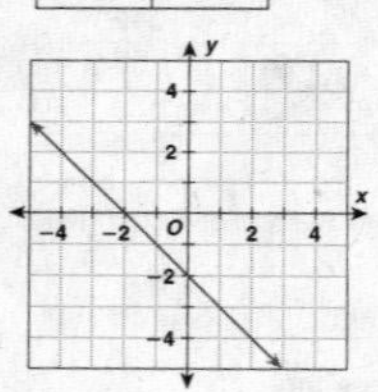

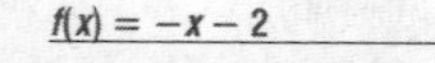

$f(x) = -x - 2$

2.

x	y
−2	4
−1	2
0	0
1	−2
2	−4

$f(x) = -2x$

Write the rule for each linear function.

3.

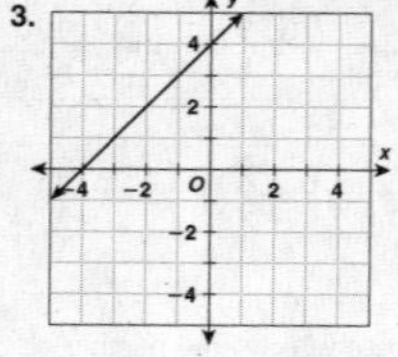

$f(x) = x + 4$

4.

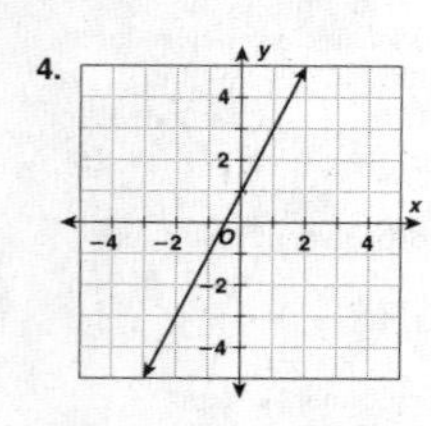

$f(x) = 2x + 1$

5. A salesperson receives a base monthly salary of $400, plus 5% of her total sales for the month. Find a rule for the linear function that describes her monthly salary. Use it to determine her salary if her total sales in January are $22,400.

   $f(x) = 0.05x + 400$, where x is the total sales for a month.

   $f(22{,}400) = 0.05 \cdot 22{,}400 + 400 = \$1{,}520$

LESSON 13-4
Practice B
Linear Functions

Determine whether each function is linear.

1. $f(x) = -3x + 2$

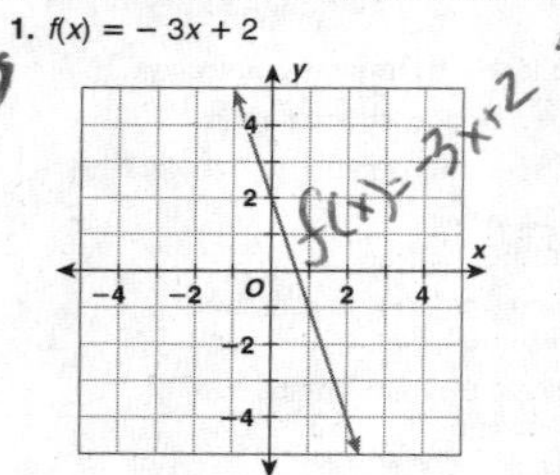

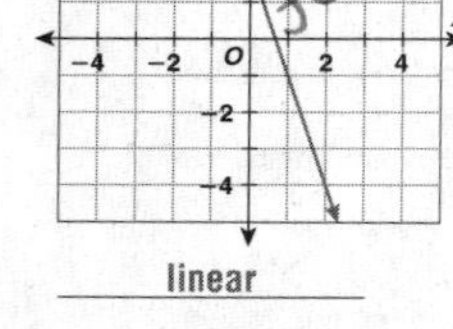

linear

2. $f(x) = x^2 - 1$

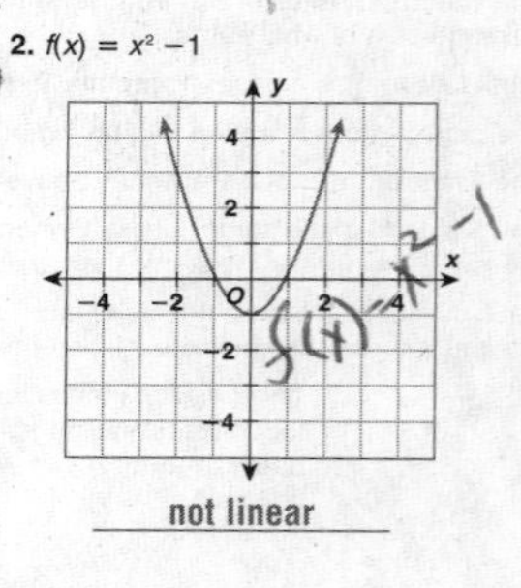

not linear

Write a rule for each linear function.

3.

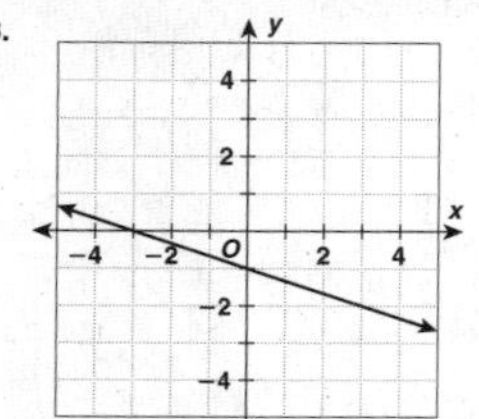

$f(x) = -\frac{1}{3}x - 1$

4.

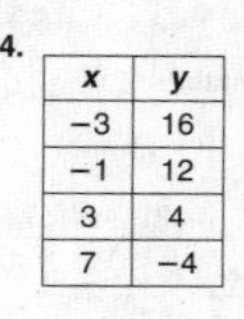

x	y
−3	16
−1	12
3	4
7	−4

$f(x) = -2x + 10$

5. At the Sweater Store, the price of a sweater is 20% more than the wholesale cost, plus a markup of $8. Find a rule for a linear function that describes the price of sweaters at the Sweater Store. Use it to determine the price of a sweater with a wholesale cost of $24.50.

   $f(x) = 1.2x + 8$, where x is the wholesale cost of the sweater.

   $f(24.50) = 1.2 \cdot 24.5 + 8 = \$37.40$

LESSON **13-4**

Practice C
Linear Functions

Write the rule for each linear function.

1.

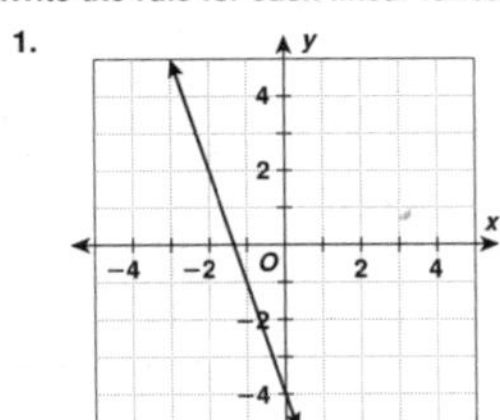

$f(x) = -3x - 4$

2.

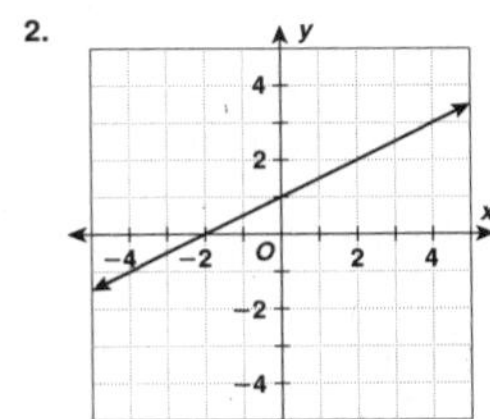

$f(x) = \frac{1}{2}x + 1$

3.

x	y
−2	2.5
−1	1.5
0	0.5
2	−2.5

$f(x) = -x + 0.5$

4.

x	y
−2	$-1\frac{1}{2}$
−1	$-1\frac{1}{4}$
0	−1
2	$-\frac{1}{2}$

$f(x) = \frac{1}{4}x - 1$

5. A cab company charges its customers a flat rate of \$10 plus \$0.75 per mile. Find a rule for the linear function that describes the cab company's basic fees. $f(x) = 0.75x + 10$

6. Use the rule for Exercise 5 to find the cab company's charge for a cab ride of 24 miles. \$28

7. A company buys a forklift truck for \$10,000. The company depreciates the truck \$125 each month for x months. Write a linear function for the forklift's value after x months. $f(x) = 10{,}000 - 125x$

8. Use the rule for Exercise 7 to find the depreciated value of the forklift truck after 15 months. \$8125

LESSON **13-4**

Reteach
Linear Functions

The graph of a **linear function** is a straight line, so you can write a **rule** for a linear function in slope-intercept form. Use function notation to show that the output value, $f(x)$, corresponds to the input value, x.

$f(x) = mx + b$ — slope, y-intercept

You can find the slope and y-intercept of a linear function in a graph of the function or in a table of its x-values and y-values.

$f(x) = 2x + 1$

x	y
−1	−1
0	1
1	3
2	5

The table shows the y-intercept, 1. $b = 1$

Substitute 1 for b into the slope-intercept form. $f(x) = mx + 1$

Substitute a pair of x- and y-values and solve. $5 = m \cdot 2 + 1$

$4 = 2m$

$2 = m$

So, the rule for the function is $f(x) = 2x + 1$.

33

Write the rule for the linear function.

5 1.

x	y
−2	−1
−1	1
0	3
1	5

$f(x) = mx + b$

$f(x) = mx + \underline{3}$

$(x, y) = (1, 5)$ $5 = m(\underline{1}) + \underline{3}$

$\underline{2} = m$

$f(x) = \underline{2x + 3}$

6 2.

x	y
−1	3
0	−1
1	−5
2	−9

$f(x) = mx + b$

$f(x) = mx \ \underline{-1}$

$(x, y) = (-1, 3)$ $\underline{3} = m(\underline{-1}) \ \underline{-1}$

$\underline{-4} = m$

$f(x) = \underline{-4x - 1}$

/11

LESSON **13-4**

Reteach
Linear Functions (continued)

If a table does not contain the y-intercept, use two points to find the slope.

Use (1, 1) and (2, 4).

$m = \frac{4-1}{2-1} = \frac{3}{1} = 3$

Then substitute the value for m and the coordinates of any point, say (1, 1), and solve for b.

Use the values of m and b to write the function rule: $f(x) = 3x - 2$

x	y
−2	−8
−1	−5
1	1
2	4

$f(x) = mx + b$
$1 = 3(1) + b$
$1 = 3 + b$
$-3 \ -3$
$-2 = b$

Write the rule for the linear function.

8 3.

x	y
−3	−17
−1	−7
1	3
3	13

Using (1, 3) and (3, 13);

3 $m = \frac{13 - 3}{3 - 1} = \frac{10}{2} = \underline{5}$

Using your value for m and (1, 3):

$f(x) = mx + b$

3 $\underline{3} = \underline{5(1)} + b$

1 $\underline{-2} = b$

1 $f(x) = \underline{5x - 2}$

Use the graph to write the function rule.

7 4. From point B, read the y-intercept.

1 $b = \underline{-3}$

Write the coordinates of two points on the line.

2 $A(\underline{2, 1})$, $C(\underline{3, 3})$

Use these points to find the slope.

3 $m = \frac{3 - 1}{3 - 2} = \frac{2}{1} = \underline{2}$

1 $f(x) = \underline{2x - 3}$

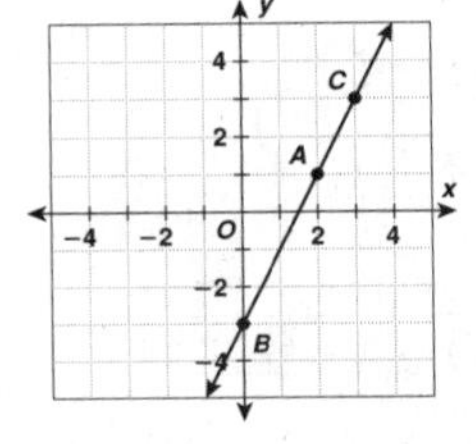

/15

LESSON **13-4**

Challenge
Function? Absolutely!

Now, you will explore a unique function, the **absolute-value function $f(x) = |x|$.**

/33

(10) 1. **a.** Complete this table of values and plot the points. 7

x	$\|x\|$	y
−3	$\|-3\|$	3
−2	$\|-2\|$	2
−1	$\|-1\|$	1
0	$\|0\|$	0
1	$\|1\|$	1
2	$\|2\|$	2
3	$\|3\|$	3

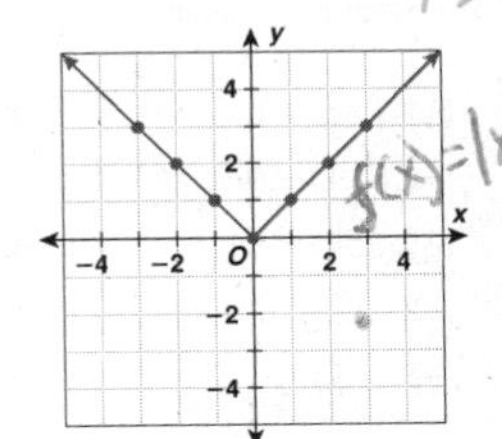

1 **b.** Your result should be in a V-formation. Connect the dots to show this.

c. Add more points to your graph that satisfy the absolute-value function $f(x) = |x|$.

2 **d.** domain? $\underline{x \text{ is any real number}}$ range? $\underline{y \geq 0}$

If you have access to a graphing instrument (calculator or computer), use it for the following exercises. If not, make tables.

(5) 2. **a.** On the same grid, draw the graphs of 2 $f(x) = |x - 1|$ and $f(x) = |x - 2|$.

b. On the same grid, draw the graphs of 2 $f(x) = |x + 1|$ and $f(x) = |x + 2|$.

c. What type of transformation takes 1 the graph of $f(x) = |x|$ to the graphs you have drawn?

translation

(4) 3. Predict the location of the vertex of the V-shape for each function that follows. (Write the coordinates.) Verify your predictions.

a. $f(x) = |x - 5|$ $\underline{(5, 0)}$

b. $f(x) = |x + 5|$ $\underline{(-5, 0)}$

c. $f(x) = |x| + 1$ $\underline{(0, 1)}$

d. $f(x) = |x| - 1$ $\underline{(0, -1)}$

/19

LESSON 13-4
Problem Solving
Linear Functions

Write the correct answer.

1. The greatest amount of snow that has ever fallen in a 24-hour period in North America was on April 14–15, 1921 in Silver Lake, Colorado. In 24 hours, 76 inches of snow fell, at an average rate of 3.2 inches per hour. Find a rule for the linear function that describes the amount of snow after x hours at the average rate.

   $f(x) = 3.2x$

2. At the average rate of snowfall from Exercise 1, how much snow had fallen in 15 hours?

   48 inches

3. The altitude of clouds in feet can be found by multiplying the difference between the temperature and the dew point by 228. If the temperature is 75°, find a rule for the linear function that describes the height of the clouds with dew point x.

   $f(x) = 228(75 - x)$

4. If the temperature is 75° and the dew point is 40°, what is the height of the clouds?

   7980 feet

For exercises 5–7, refer to the table below, which shows the relationship between the number of times a cricket chirps in a minute and temperature.

Cricket Chirps/min	Temperature (°F)
80	60
100	65
120	70
140	75

5. Find a rule for the linear function that describes the temperature based on x, the number of cricket chirps in a minute based on temperature.
   - A $f(x) = x + 5$
   - **(B)** $f(x) = \frac{x}{4} + 40$
   - C $f(x) = x - 20$
   - D $f(x) = \frac{x}{2} + 20$

6. What is the temperature if a cricket chirps 150 times in a minute?
   - **(F)** 77.5°F
   - G 95°F
   - H 130°F
   - J 155°F

7. If the temperature is 85°F, how many times will a cricket chirp in a minute?
   - A 61
   - B 105
   - **(C)** 180
   - D 200

LESSON 13-4
Reading Strategies
Use a Graphic Aid

The graph of a **linear function** is a straight line, so you can write a **rule** for a linear function in slope-intercept form.

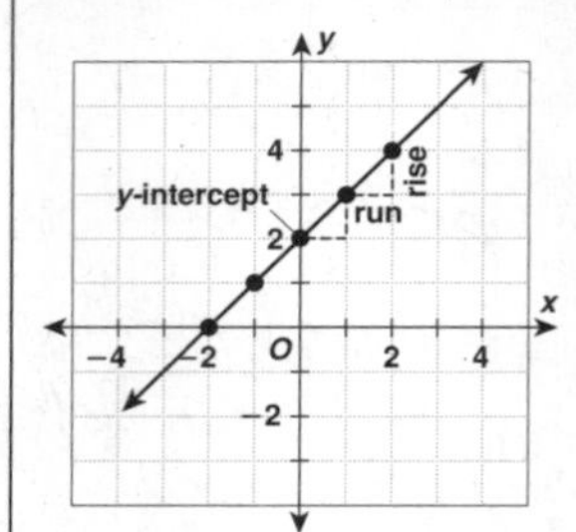

$f(x) = mx + b$

slope intercept

Use the graph to answer each question.

1. What type of function is pictured in the graph? **linear function**
2. What is the name for the y-coordinate of the point where the line crosses the y-axis? **the y-intercept**
3. What is the y-intercept for this graph? **2**

You can use the graph to find the slope of the line. Find the "rise over run," or the change in y to the change in x.

$$\text{slope} = \frac{\text{rise}}{\text{run}} = \frac{\text{change in } y}{\text{change in } x}$$

Answer each question.

4. How can you find the slope of a linear function?

   Sample answer: Write the ratio of the change in y to the change in x of any two points on the graph of the function.

5. Find the slope of the graph above. **$\frac{1}{1}$, or 1**
6. Complete to write a rule for the function. $f(x) =$ **1** $x +$ **2**

LESSON 13-4
Puzzles, Twisters and Teasers
Bee-lieve It or Not!

Fill in the missing coordinate for each point on the graph. Use the corresponding letter to solve the riddle. (Hint: There is one letter that you will not need.)

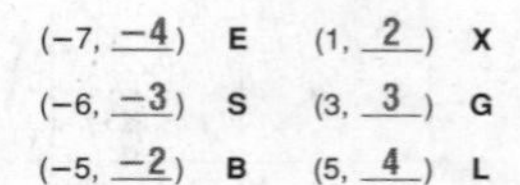

(−7, **−4**)	E	(1, **2**)	X
(−6, **−3**)	S	(3, **3**)	G
(−5, **−2**)	B	(5, **4**)	L
(−3, **−1**)	P	(6, **5**)	N
(−1, **1**)	A	(7, **6**)	I

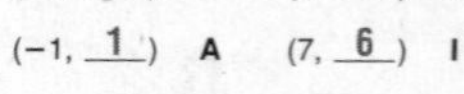

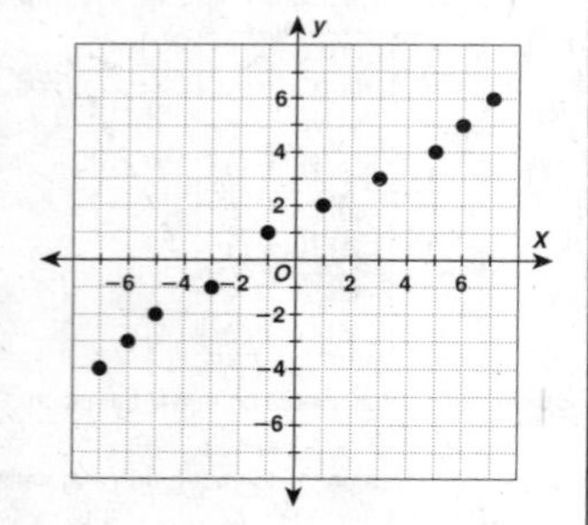

What is more amazing than a talking dog?

A
1

S **P** **E** **L** L **I** **N** **G**
−3 −1 −4 4 6 5 3

B E **E**
−2 −4

LESSON 13-5
Practice A
Exponential Functions

Complete the table for each exponential functions.

1. $f(x) = 2^x$

x	y
−2	$y = 2^{-2} = \frac{1}{4}$
−1	$y = 2^{-1} = \frac{1}{2}$
0	$y = 2^0 = 1$
1	$y = 2^1 = 2$
2	$y = 2^2 = 4$

2. $f(x) = (0.2)4^x$

x	y
−2	$y = (0.2)4^{-2} = 0.0125$
−1	$y = (0.2)4^{-1} = 0.05$
0	$y = (0.2)4^0 = 0.2$
1	$y = (0.2)4^1 = 0.8$
2	$y = (0.2)4^2 = 3.2$

Create a table for each exponential function, and use it to graph the function.

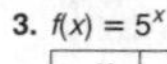

3. $f(x) = 5^x$

x	y
−2	$y = 5^{-2} = \frac{1}{25}$
−1	$y = 5^{-1} = \frac{1}{5}$
0	$y = 5^0 = 1$
1	$y = 5^1 = 5$

4. $f(x) = 4 \cdot 2^x$

x	y
−2	$y = 4 \cdot 2^{-2} = 1$
−1	$y = 4 \cdot 2^{-1} = 2$
0	$y = 4 \cdot 2^0 = 4$
1	$y = 4 \cdot 2^1 = 8$

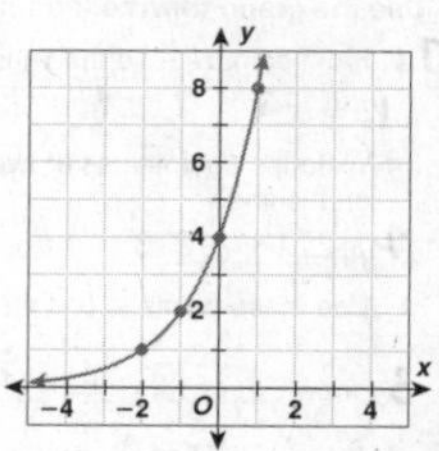

5. The exponential function $f(x) = 1000 \cdot 1.03^x$ describes the increase in a bank deposit of \$1000 with a 3% annual interest rate for x years. Find the value of the deposit after 2 years. **\$1060.90**

LESSON **13-5**

Practice B

Exponential Functions

Create a table for each exponential function, and use it to graph the function.

1. $f(x) = 0.5 \cdot 4^x$

x	y
−1	$y = 0.5 \cdot 4^{-1} = 0.125$
0	$y = 0.5 \cdot 4^0 = 0.5$
1	$y = 0.5 \cdot 4^1 = 2$
2	$y = 0.5 \cdot 4^2 = 8$

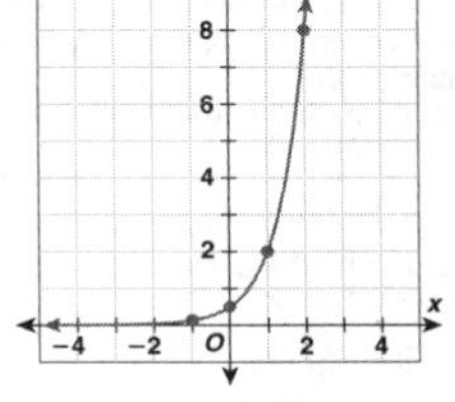

2. $f(x) = \frac{1}{3} \cdot 3^x$

x	y
−1	$y = \frac{1}{3} \cdot 3^{-1} = \frac{1}{9}$
0	$y = \frac{1}{3} \cdot 3^0 = \frac{1}{3}$
1	$y = \frac{1}{3} \cdot 3^1 = 1$
2	$y = \frac{1}{3} \cdot 3^2 = 3$

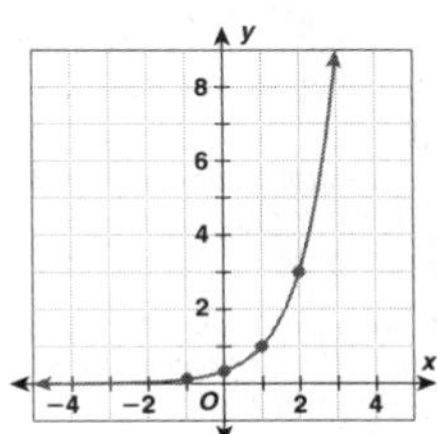

3. A forestry department introduce 500 fish to a lake. The fish are expected to increase at a rate of 35% each year. Write an exponential function to calculate the number of fish in the lake at the end of each year. Predict how many fish will be in the lake at the end of 5 years. $f(x) = 500\,(1.35)^x$; 2242 fish

4. A stock valued at \$756 has been declining steadily at the rate of 4% a year for the last few years. If this decline continues, predict what the value of the stock will be at the end of 3 years. \$668.86

5. Todd's starting salary at his new job is \$400 a week. He is promised a 3% increase in salary every year. Predict to the nearest dollar what Todd's expected yearly salary will be after working for 4 years. \$23,411

LESSON **13-5**

Practice C

Exponential Functions

1. For each exponential function, find $f(3)$, $f(0)$, $f(-4)$.

	f(3)	f(0)	f(−4)
$f(x) = 4^x$	64	1	$\frac{1}{256}$
$f(x) = 0.8^x$	0.512	1	2.44140625
$f(x) = 15^x$	3375	1	$\frac{1}{50{,}625}$
$f(x) = 75 \cdot \left(\frac{1}{4}\right)^x$	$1\frac{11}{64}$	75	19,200

Write the equation of the exponential function that passes through the given points. Use the form $f(x) = p \cdot a^x$.

2. (0, 2) and (1, 8) $f(x) = 2 \cdot 4^x$

3. (0, 5) and (1, 10) $f(x) = 5 \cdot 2^x$

4. (0, 6) and (1, 3) $f(x) = 6 \cdot \left(\frac{1}{2}\right)^x$

Graph the exponential function of the form $f(x) = p \cdot a^x$.

5. $p = 4$, $a = 2$

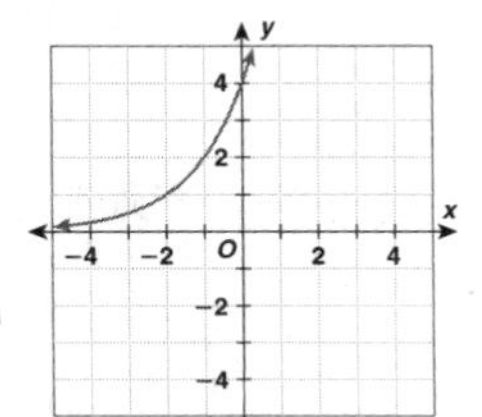

6. $p = -2$, $a = \frac{1}{2}$

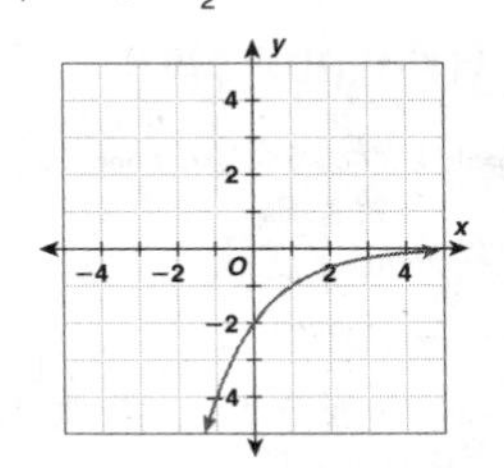

7. What percent of decrease is modeled in the exponential function $f(x) = 100 \cdot 0.92^x$? 8%

8. Mr. Harry has a rabbit farm with 10 rabbits. If the number of rabbits triples each half-year, how many rabbits will be on the farm after 3 years? 7290 rabbits

LESSON **13-5**

Reteach

Exponential Functions

A function that has the input value x in the exponent is called an **exponential function.** The base number a is positive. $f(x) = a^x$

Situation 1 $a > 1$, say $a = 3$ — $f(x) = 3^x$

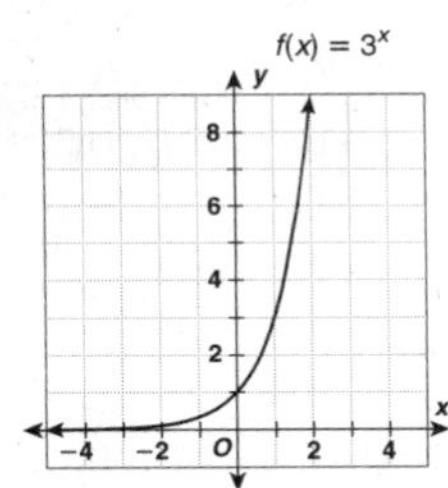

This graph rises from left to right.

Situation 2 $a < 1$, say $a = \frac{1}{3}$ — $f(x) = \left(\frac{1}{3}\right)^x$

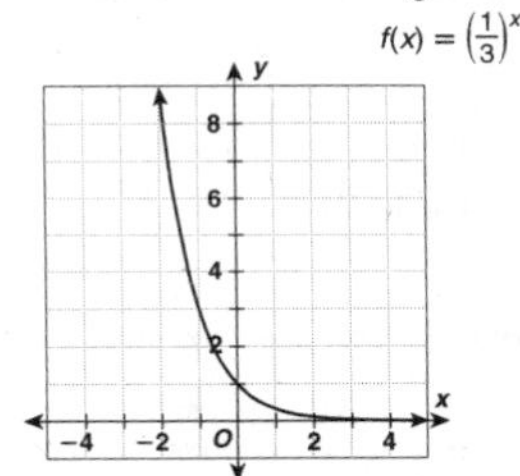

This graph falls from left to right.

For both situations: The domain is the set of all real numbers.
The range is the set of positive real numbers.
The y-intercept is 1.

Complete the table for each function.
Graph both functions on the same grid. Label each function.

1. $f(x) = 2^x$

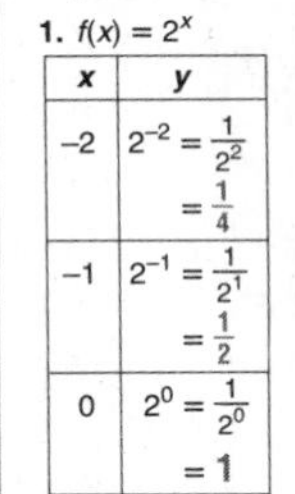

x	y
−2	$2^{-2} = \frac{1}{2^2} = \frac{1}{4}$
−1	$2^{-1} = \frac{1}{2^1} = \frac{1}{2}$
0	$2^0 = \frac{1}{2^0} = 1$
1	$2^1 = 2$
2	$2^2 = 4$

2. $f(x) = \left(\frac{1}{2}\right)^x$

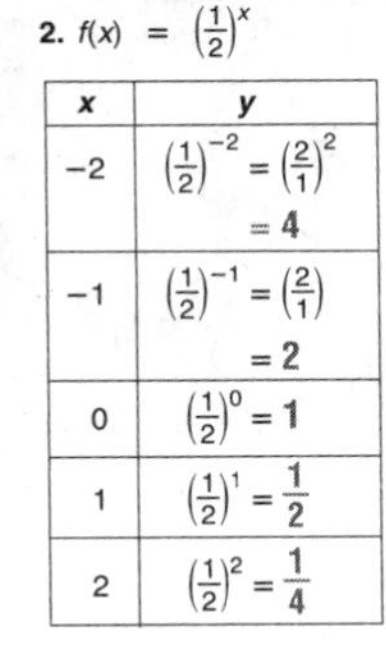

x	y
−2	$\left(\frac{1}{2}\right)^{-2} = \left(\frac{2}{1}\right)^2 = 4$
−1	$\left(\frac{1}{2}\right)^{-1} = \left(\frac{2}{1}\right) = 2$
0	$\left(\frac{1}{2}\right)^0 = 1$
1	$\left(\frac{1}{2}\right)^1 = \frac{1}{2}$
2	$\left(\frac{1}{2}\right)^2 = \frac{1}{4}$

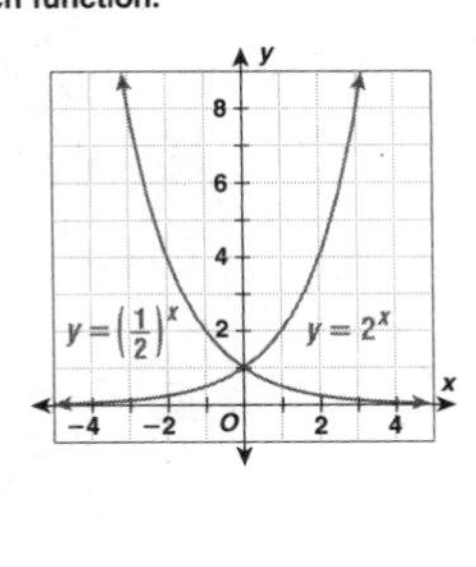

LESSON **13-5**

Reteach

Exponential Functions (continued)

When an exponential function has a **starting number** p, its rule can be written $f(x) = p \cdot a^x$.

If $a > 1$, the output grows as the input grows, and the function is called an **exponential growth function.** If $a < 1$, the output shrinks as the input grows, and the function is called an **exponential decay function.**

Say a bacteria doubles in number every 10 hours, and there are 100 bacteria to begin with. How many bacteria will there be after 50 hours?

$a = 2$ Find the common ratio.
$p = 100$ Find the starting number.
$x = 5$ Find the number of 10-hour periods in 50 hours.
$f(x) = p \cdot a^x$
$f(x) = 100 \cdot 2^5$ Substitute into the function rule.
$f(x) = 3200$

So, after 50 hours, there will be 3200 bacteria.

Write and apply an exponential growth function.

3. Consider a bacteria that triples in number every 24 hours. $a =$ 3

Suppose there are 40 bacteria to begin. $p =$ 40

Write a function that models the number of bacteria present after x 24-hour periods. $f(x) = 40 \cdot 3^x$

Use this function to predict the number of bacteria present after 48 hours. $x =$ 2

$f(x) = 40 \cdot 3^2 = 360$

So, after 48 hours, there will be 360 bacteria present.

4. The half-life of Barium-131 is 12 days, which means it takes 12 days for half of the substance to decompose. $a = \frac{1}{2}$

There are 100 grams to begin. $p =$ 100

Write a function that models the number of grams present after x 12-day periods. $f(x) = 100 \cdot \left(\frac{1}{2}\right)^x$

Use this function to predict the number of grams present after 36 days. $x =$ 3

$f(x) = 100 \cdot \left(\frac{1}{2}\right)^3 = \frac{100}{8} = 12.5$

So, after 36 days, there will be 12.5 grams of Barium-131 remaining.

LESSON 13-5

Challenge

Exponent Search

Now, you will see how to solve some exponential equations.

Solve for x:	$2^{x+1} = 16$	an exponential equation
	$2^{x+1} = 2^4$	Write both sides as powers of the same base.
	$x + 1 = 4$	Equate the exponents.
	$-1 \quad -1$; $x = 3$	Solve the linear equation for x.
Check:	$2^{x+1} = 16$	Use the original equation.
	$2^{3+1} \stackrel{?}{=} 16$	Substitute $x = 3$.
	$2^4 \stackrel{?}{=} 16$	
	$16 = 16$ ✓	

Solve and check each exponential equation.

1. $2^{x+3} = 64$

$2^{x+3} = 2^6$

$x + 3 = 6$

$x = 3$

Check: $2^{3+3} \stackrel{?}{=} 64$; $2^6 \stackrel{?}{=} 64$; $64 = 64$ ✓

2. $3^{x+2} = 9^x$

$3^{x+2} = (3^2)^x$

$3^{x+2} = 3^{2x}$

$x + 2 = 2x$

$x = 2$

Check: $3^{2+2} \stackrel{?}{=} 9^2$; $3^4 \stackrel{?}{=} 9^2$; $81 = 81$ ✓

3. $4^{2x+1} = 8^{2x}$

$(2^2)^{2x+1} = (2^3)^{2x}$

$2^{4x+2} = 2^{6x}$

$4x + 2 = 6x$

$x = 1$

Check: $4^{2(1)+1} \stackrel{?}{=} 8^{2(1)}$; $4^3 \stackrel{?}{=} 8^2$; $64 = 64$ ✓

4. $27^x = 9^{2x-1}$

$(3^3)^x = (3^2)^{2x-1}$

$3^{3x} = 3^{4x-2}$

$3x = 4x - 2$

$x = 2$

Check: $27^2 \stackrel{?}{=} 9^{2(2)-1}$; $729 = 729$ ✓

5. $\left(\frac{1}{2}\right)^x = 4$

$(2^{-1})^x = 2^2$

$2^{-x} = 2^2$

$-x = 2$

$x = -2$

Check: $\left(\frac{1}{2}\right)^{-2} \stackrel{?}{=} 4$; $4 = 4$ ✓

6. $\left(\frac{1}{3}\right)^{x+1} = 27$

$(3^{-1})^{x+1} = 3^3$

$3^{-x-1} = 3^3$

$-x - 1 = 3$

$x = -4$

Check: $\left(\frac{1}{3}\right)^{-3} \stackrel{?}{=} 27$; $27 = 27$ ✓

LESSON 13-5

Problem Solving

Exponential Functions

From 1950 to 2000, the world's population grew exponentially. The function that models the growth is $f(x) = 1.056 \cdot 1.018^x$ where x is the year ($x = 50$ represents 1950) and $f(x)$ is the population in billions. Round each number to the nearest hundredth.

1. Estimate the world's population in 1950.
   2.58 billion
2. Estimate the world's population in 2005.
   6.87 billion
3. Predict the world's population in 2025.
   9.82 billion
4. Predict the world's population in 2050.
   15.34 billion

Insulin is used to treat people with diabetes. The table below shows the percent of an insulin dose left in the body at different times after injection.

Time elapsed (min)	Percent remaining
0	100
48	50
96	25
144	12.5

5. Which ordered pair does not represent a half-life of insulin?
   A (24, 70.71) C (48, 50)
   (B) (50, 50) D (72, 35.35)
6. Write an exponential function that describes the percent of insulin in the body after x half-lives.
   (F) $f(x) = 100\left(\frac{1}{2}\right)^x$ H $f(x) = 2(100)^x$
   G $f(x) = 10\left(\frac{1}{2}\right)^x$ J $f(x) = 48\left(\frac{1}{2}\right)^x$
7. What percent of insulin would be left in the body after 6 hours?
   A 0.25% **(C)** 0.55%
   B 0.39% D 1.56%
8. What percent of insulin would be left in the body after 9 hours?
   (F) 0.04% H 0.17%
   G 0.12% J 0.26%
9. A new form of insulin that is being developed has a half-life of 9 hours. Write an exponential function that describes the percent of insulin in the body after x half-lives.
   (A) $f(x) = 100\left(\frac{1}{2}\right)^x$ C $f(x) = 2(100)^x$
   B $f(x) = 9\left(\frac{1}{2}\right)^x$ D $f(x) = 100(9)^x$
10. What percent of the new form of insulin would be left in the body after 9 hours?
    F 12.5% **(H)** 50%
    G 25% J 75%

LESSON 13-5

Reading Strategies

Use a Context

Will started a savings plan. He saved \$2 the first month. His goal was to save twice as much money each month.

Month	1	2	3	4
Amount Saved	\$2	\$4	\$8	\$16

Use the table to answer each question.

1. How much money did Will save the second month? **\$4**
2. How many times as great was the second month's savings as the first month's savings? **2 times**
3. How much money did Will save in the third month? **\$8**
4. How many times as great was the third month's savings as the first month's savings? **4 times**

Will noticed a pattern in his savings and made another table.

Month	Pattern	Savings
x	2^x	y
1	2^1	\$2
2	2^2	\$4
3	2^3	\$8
4	2^4	\$16

The function $f(x) = 2^x$ is an example of an exponential function.

Answer each question.

5. What does x stand for in this exponential function? **the month**
6. What does $f(x)$ stand for in this exponential function?
   the amount saved that month

LESSON 13-5

Puzzles, Twisters and Teasers

Collar I.D.?

Complete the chart which shows a part of the U.S. population that is growing exponentially. Each answer has a corresponding letter. Use the corresponding letters to solve the riddle.

Americans over 100 Years Old (thousands)										
Year	**2000**	2010	**2020**	2030	**2040**	2050	**2060**	2070	2080	2090
Population	70	**140**	280	**560**	1120	**2240**	4480	**8960**	**17920**	**35840**
	R	G	E	O	V	L	I	D	C	N

What do you get when you cross a dog and a phone?

A **G** (140) **O** (560) **L** (2240) **D** (8960) **E** (2020) **N** (35840)

R (2000) E **C** (17920) E **I** (2060) **V** (2040) E R

LESSON 13-6 Practice A
Quadratic Functions

Complete the table for each quadratic function.

1. $f(x) = x^2 - 1$

x	$f(x) = x^2 - 1$
−3	$f(-3) = (-3)^2 - 1 = 8$
−2	$f(-2) = (-2)^2 - 1 = 3$
−1	$f(-1) = (-1)^2 - 1 = 0$
0	$f(0) = (0)^2 - 1 = -1$
1	$f(1) = (1)^2 - 1 = 0$
2	$f(2) = (2)^2 - 1 = 3$
3	$f(3) = (3)^2 - 1 = 8$

2. $f(x) = x^2 - 2x + 3$

x	$f(x) = x^2 - 2x + 3$
−3	$f(x) = (-3)^2 - 2(-3) + 3 = 18$
−2	$f(x) = (-2)^2 - 2(-2) + 3 = 11$
−1	$f(x) = (-1)^2 - 2(-1) + 3 = 6$
0	$f(x) = (0)^2 - 2(0) + 3 = 3$
1	$f(x) = (1)^2 - 2(1) + 3 = 2$
2	$f(x) = (2)^2 - 2(2) + 3 = 3$
3	$f(x) = (3)^2 - 2(3) + 3 = 6$

3. Complete the table and graph the quadratic function, $f(x) = x^2 + 4x + 2$.

x	$f(x) = x^2 + 4x + 2$
−5	$f(-5) = (-5)^2 + 4(-5) + 2 = 7$
−4	$f(-4) = (-4)^2 + 4(-4) + 2 = 2$
−3	$f(-3) = (-3)^2 + 4(-3) + 2 = -1$
−2	$f(-2) = (-2)^2 + 4(-2) + 2 = -2$
−1	$f(-1) = (-1)^2 + 4(-1) + 2 = -1$
0	$f(0) = (0)^2 + 4(0) + 2 = 2$
1	$f(1) = (1)^2 + 4(1) + 2 = 7$

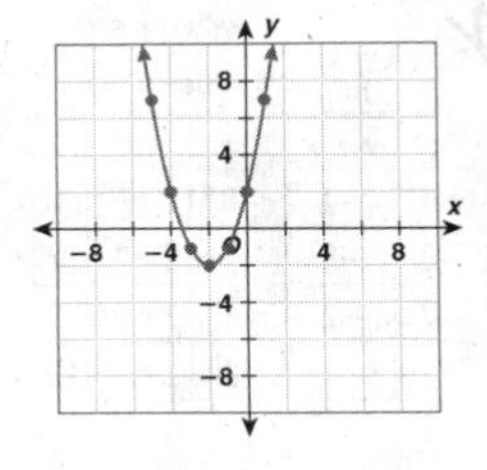

4. One number is 6 greater than another number. Their product is given by the function $f(x) = x^2 + 6x$. Which pair of numbers results in the least product?

3 and −3

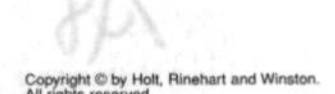

LESSON 13-6 Practice B
Quadratic Functions

Create a table for each quadratic function, and use it to make a graph.

1. $f(x) = x^2 - 5$

x	$f(x) = x^2 - 5$
−3	$f(-3) = (-3)^2 - 5 = 4$
−1	$f(-1) = (-1)^2 - 5 = -4$
0	$f(0) = (0)^2 - 5 = -5$
2	$f(2) = (2)^2 - 5 = -1$
3	$f(3) = (3)^2 - 5 = 4$

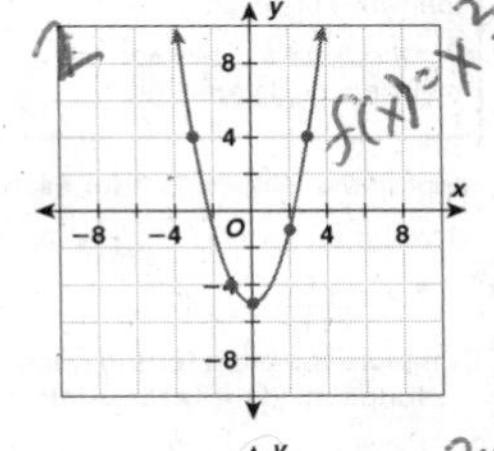

2. $f(x) = x^2 - 2x + 3$

x	$f(x) = x^2 - 2x + 3$
3	$f(3) = (3)^2 - 2(3) + 3 = 6$
2	$f(2) = (2)^2 - 2(2) + 3 = 3$
1	$f(1) = (1)^2 - 2(1) + 3 = 2$
0	$f(0) = (0)^2 - 2(0) + 3 = 3$
−1	$f(-1) = (-1)^2 - 2(-1) + 3 = 6$

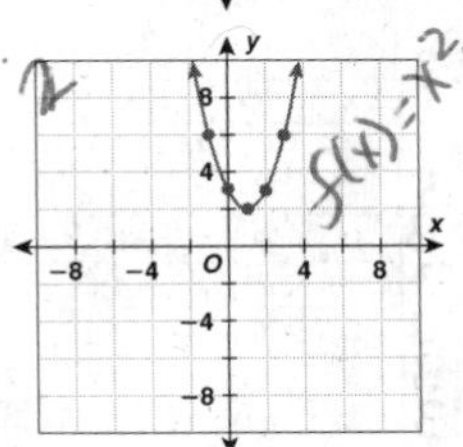

3. Find $f(-3)$, $f(0)$, $f(3)$ for each quadratic function.

	$f(-3)$	$f(0)$	$f(3)$
$f(x) = x^2 - 2x + 1$	16	1	4
$f(x) = x^2 - 6$	3	−6	3
$f(x) = x^2 - x + 3$	15	3	9

4. The function $f(t) = -4.9t^2$ gives the distance in meters that an object will fall toward Earth in t seconds. Find the distance an object will fall in 1, 2, 3, 4, and 5 seconds. (Note that the distance traveled by a falling object is shown by a negative number.)

4.9 m, 19.6 m, 44.1 m, 78.4 m, and 122.5 m

LESSON 13-6 Practice C
Quadratic Functions

Create a table for each quadratic function, and use it to make a graph.

1. $f(x) = x^2 - 5x + 6$

x	$f(x) = x^2 - 5x + 6$
−1	$f(-1) = (-1)^2 - 5(-1) + 6 = 12$
0	$f(0) = 0^2 - 5(0) + 6 = 6$
1	$f(1) = 1^2 - 5(1) + 6 = 2$
2	$f(2) = 2^2 - 5(2) + 6 = 0$
3	$f(3) = 3^2 - 5(3) + 6 = 0$
4	$f(4) = 4^2 - 5(4) + 6 = 2$
5	$f(5) = 5^2 - 5(5) + 6 = 6$

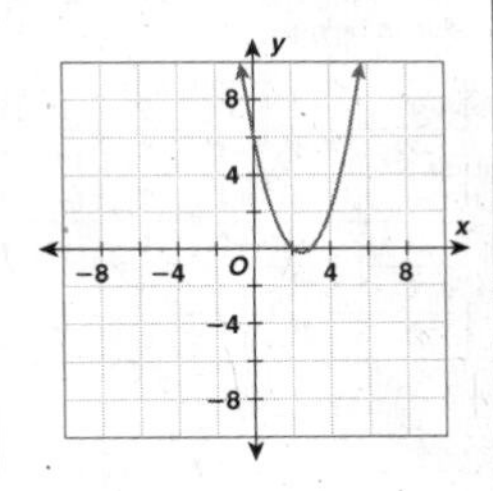

2. $f(x) = (x - 2)(x - 4)$

x	$f(x) = (x - 2)(x - 4)$
0	$f(0) = (0 - 2)(0 - 4) = 8$
1	$f(1) = (1 - 2)(1 - 4) = 3$
2	$f(2) = (2 - 2)(2 - 4) = 0$
3	$f(3) = (3 - 2)(3 - 4) = -1$
4	$f(4) = (4 - 2)(4 - 4) = 0$
5	$f(5) = (5 - 2)(5 - 4) = 3$

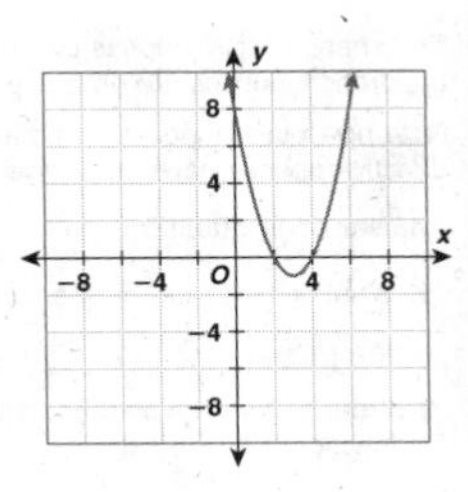

Find $f(-5)$, $f(0)$, $f(5)$ for each quadratic function.

3. $f(x) = x^2 - x - 9$

$f(-5) = 21$, $f(0) = -9$, $f(5) = 11$

3. $f(x) = (x - 9)(x + 8)$

$f(-5) = -42$, $f(0) = -72$, $f(5) = -52$

4. The height in meters of a ball thrown at a certain speed is given by the function $f(t) = -4.9t^2 + 19.6t$, where t is the elapsed time in seconds. Find the height of the ball after 1, 2, 3, and 4 seconds.

14.7 m, 19.6 m, 14.7 m, and 0 m

LESSON 13-6 Reteach
Quadratic Functions

A **quadratic function** has a variable that is squared.

general quadratic function $f(x) = ax^2 + bx + c$

square term (ax^2) ; y-intercept (c)

$f(x) = x^2 - 4x + 3$

The graph of a quadratic function is a **parabola**, a curve that falls on one side of a turning point and rises on the other. You can make a table of a function's values and use them to graph the function.

x	$f(x) = x^2 - 4x + 3$
−1	$f(-1) = (-1)^2 - 4(-1) + 3 = 8$
0	$f(0) = 0^2 - 4(0) + 3 = 3$
1	$f(1) = 1^2 - 4(1) + 3 = 0$
2	$f(2) = 2^2 - 4(2) + 3 = -1$
3	$f(3) = 3^2 - 4(3) + 3 = 0$
4	$f(4) = 4^2 - 4(4) + 3 = 3$
5	$f(5) = 5^2 - 4(5) + 3 = 8$

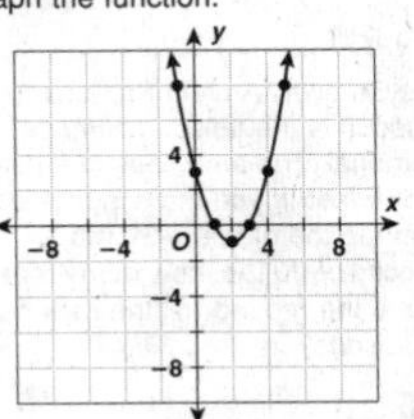

Complete the table for the quadratic function and use it to graph the function.

1. $f(x) = x^2 - 2x - 3$

x	$f(x) = x^2 - 2x - 3$
−2	$f(-2) = (-2)^2 - 2(-2) - 3 = 5$
−1	$f(-1) = (-1)^2 - 2(-1) - 3 = 0$
0	$f(0) = (0)^2 - 2(0) - 3 = -3$
1	$f(1) = (1)^2 - 2(1) - 3 = -4$
2	$f(2) = (2)^2 - 2(2) - 3 = -3$
3	$f(3) = (3)^2 - 2(3) - 3 = 0$
4	$f(4) = (4)^2 - 2(4) - 3 = 5$

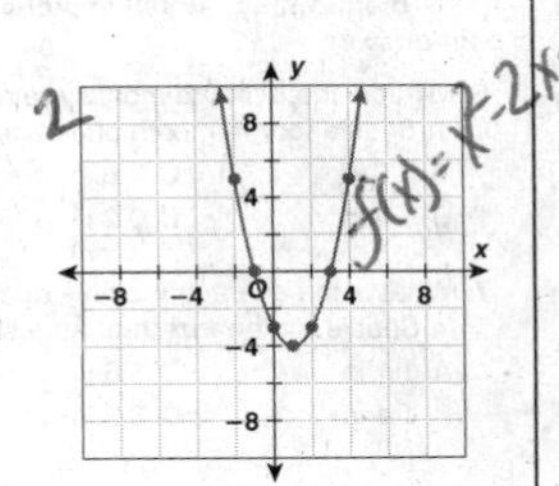

LESSON **13-6** **Reteach**

Quadratic Functions (continued)

When a quadratic function is written as the product of two differences, you can read the two x-intercepts.

$f(x) = (x - r)(x - s)$ — r and s: x-intercepts

For the quadratic function $f(x) = (x - 3)(x + 4)$, the x-intercepts are 3 and -4.

Identify the x-intercepts for each function.

2. $f(x) = (x - 4)(x - 7)$
   4 and 7
3. $f(x) = (x + 1)(x - 5)$
   -1 and 5
4. $f(x) = (x + 2)(x + 4)$
   -2 and -4

Complete the table for the quadratic function and use it to graph the function. Identify the x-intercepts and the y-intercept.

5. $f(x) = (x + 1)(x - 3)$

x	$f(x) = (x + 1)(x - 3)$
-2	$f(-2) = (-2 + 1)(-2 - 3) = 5$
-1	$f(-1) = (-1 + 1)(-1 - 3) = 0$
0	$f(0) = (0 + 1)(0 - 3) = -3$
1	$f(1) = (1 + 1)(1 - 3) = -4$
2	$f(2) = (2 + 1)(2 - 3) = -3$
3	$f(3) = (3 + 1)(3 - 3) = 0$
4	$f(4) = (4 + 1)(4 - 3) = 5$

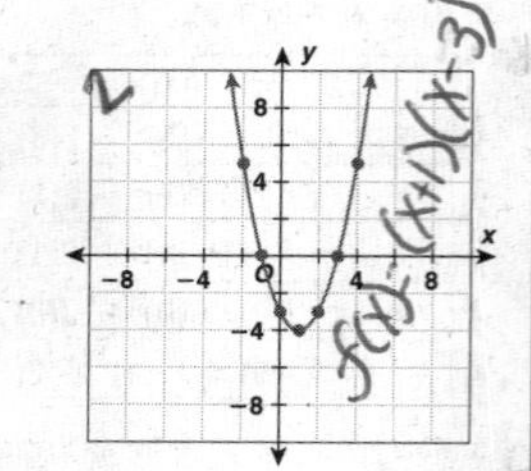

x-intercepts = -1 and 3

y-intercept = -3

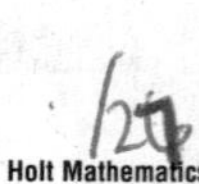
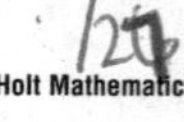

LESSON **13-6** **Challenge**

A Piece of This, a Piece of That

A function defined differently over various parts of its domain is called a **piecewise function.**

$$f(x) = \begin{cases} x - 2 & \text{when } x < 0 \\ x^2 & \text{when } x \geq 0 \end{cases}$$

This function consists of the line $y = x - 2$ when x is negative and the parabola $y = x^2$ when x is nonnegative.

x	$x - 2$	y		x	x^2	y
-4	$-4 - 2$	-6		0	0^2	0
-3	$-3 - 2$	-5		1	1^2	1
-2	$-2 - 2$	-4		2	2^2	4
-1	$-1 - 2$	-3		3	3^2	9

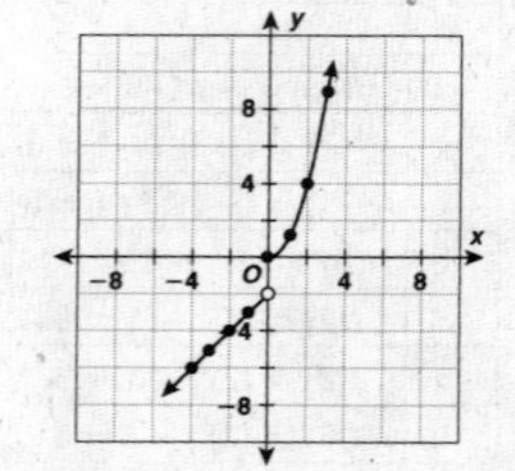

Note the open hole on the line at $(0, -2)$.
When $x = 0$, the point on this graph is on the parabola, not on the line.

Graph each piecewise function.

1. $f(x) = \begin{cases} x & \text{when } x < 0 \\ 2x^2 & \text{when } x \geq 0 \end{cases}$

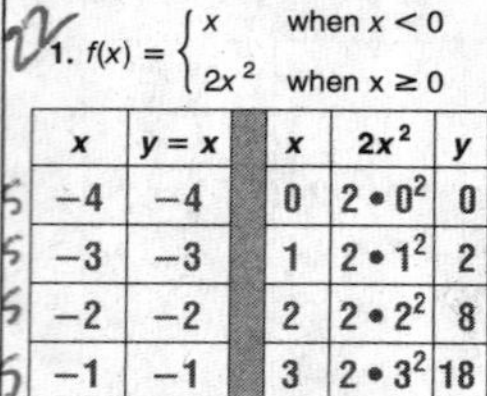

x	$y = x$		x	$2x^2$	y
-4	-4		0	$2 \bullet 0^2$	0
-3	-3		1	$2 \bullet 1^2$	2
-2	-2		2	$2 \bullet 2^2$	8
-1	-1		3	$2 \bullet 3^2$	18

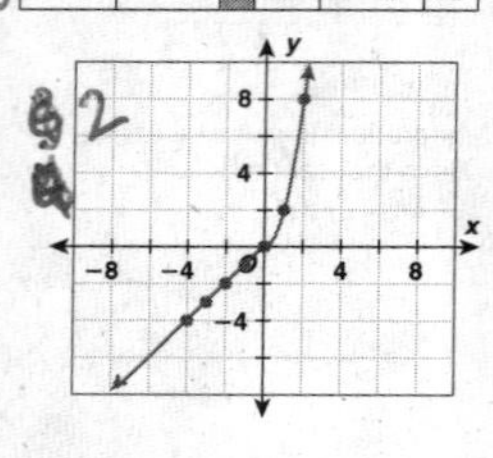

2. $f(x) = \begin{cases} x + 2 & \text{when } x < 0 \\ -x^2 & \text{when } x \geq 0 \end{cases}$

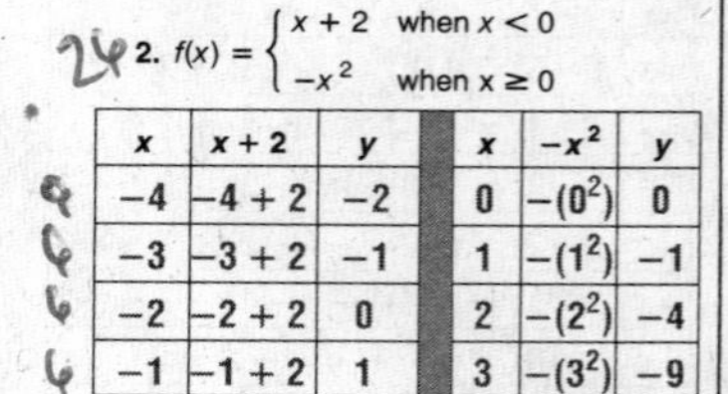

x	$x + 2$	y		x	$-x^2$	y
-4	$-4 + 2$	-2		0	$-(0^2)$	0
-3	$-3 + 2$	-1		1	$-(1^2)$	-1
-2	$-2 + 2$	0		2	$-(2^2)$	-4
-1	$-1 + 2$	1		3	$-(3^2)$	-9

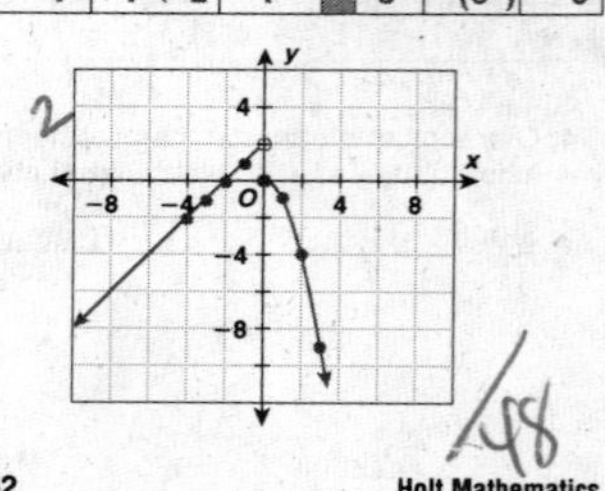

LESSON **13-6** **Problem Solving**

Quadratic Functions

To find the time it takes an object to fall, you can use the equation $h = -16t^2 - vt + s$ where h is the height in feet, t is the time in seconds, v is the initial velocity, and s is the starting height in feet. Write the correct answer.

1. If a construction worker drops a tool from 240 feet above the ground, how many feet above the ground will it be in 2 seconds? Hint: $v = 0$, $s = 240$.

   176 feet

2. How long will it take the tool in Exercise 1 to hit the ground? Round to the nearest hundredth.

   3.87 seconds

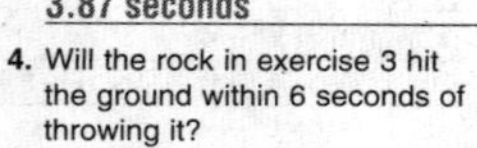

3. The Gateway Arch in St. Louis, Missouri is the tallest manmade memorial. The arch rises to a height of 630 feet. If you throw a rock down from the top of the arch with a velocity of 20 ft/s, how many feet above the ground will the rock be in 2 seconds?

   526 feet

4. Will the rock in exercise 3 hit the ground within 6 seconds of throwing it?

   yes

The average monthly rainfall for Seattle, Washington can be approximated by the equation $f(x) = 0.147x^2 - 1.890x + 7.139$ where x is the month (January: $x = 1$, February, $x = 2$, etc.) and $f(x)$ is the monthly rainfall in inches. Choose the letter for the best answer.

5. What is the average monthly rainfall in Seattle for the month of January?
   A 3.7 in
   (B) 5.4 in
   C 7.6 in
   D 9.2 in

6. What is the average monthly rainfall in Seattle for the month of April?
   F 0.2 in
   G 1.4 in
   (H) 1.9 in
   J 2.8 in

7. What is the average monthly rainfall in Seattle for the month of August?
   A 1.1 in
   (B) 1.4 in
   C 5.6 in
   D 6.8 in

8. In what month does it rain the least in Seattle, Washington?
   F May
   (G) June
   H July
   J August

LESSON **13-6** **Reading Strategies**

Analyze Information

A **quadratic function** contains a variable to the second power.

This equation is a quadratic function. → $y = x^2 + 1$

A table of values and the graph for this quadratic function are shown below.

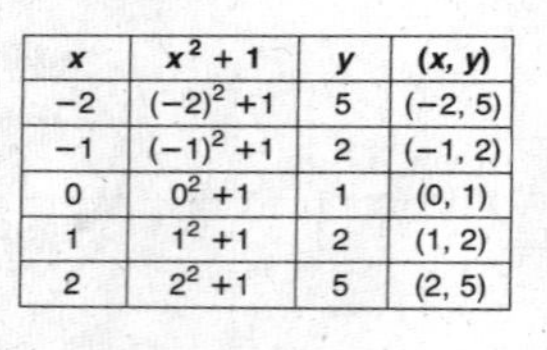

x	$x^2 + 1$	y	(x, y)
-2	$(-2)^2 + 1$	5	$(-2, 5)$
-1	$(-1)^2 + 1$	2	$(-1, 2)$
0	$0^2 + 1$	1	$(0, 1)$
1	$1^2 + 1$	2	$(1, 2)$
2	$2^2 + 1$	5	$(2, 5)$

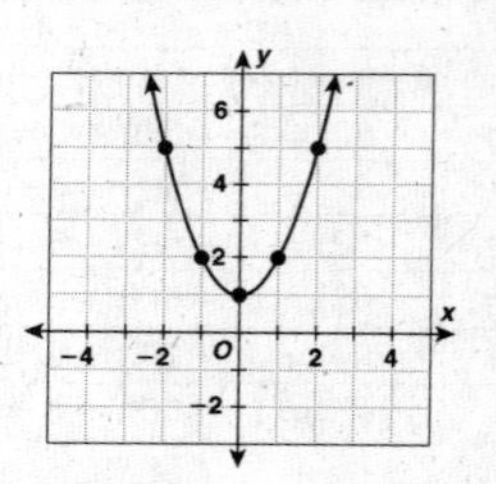

The shape of this graph is called a **parabola.** The graphs of quadratic functions are always parabolas.

Note how pairs of points that make up the parabola are located directly opposite each other on the graph.

Answer each question.

1. What is the power of one variable in a quadratic function?

   the power of 2

2. What name is given to the shape of the graph for a quadratic function?

   a parabola

3. In your own words, how would you describe the shape of the graph of a quadratic function?

   Possible answer: a U-shaped graph

4. The coordinates $(-1, 2)$ identify a point on the left side of the parabola. Write the coordinates that identify the opposite point on the graph.

   $(1, 2)$

LESSON 13-6

Puzzles, Twisters and Teasers

Paws for the Cause!

Solve each equation. Then plot the points from the table on the graph and connect them with a smooth curve. Each point has a corresponding letter. Use the letters to solve the riddle.

x	$f(x) = x^2 - 2$	y
−3	$(-3)^2 - 2$	7
−2	$(-2)^2 - 2$	2
−1	$(-1)^2 - 2$	−1
0	$(0)^2 - 2$	−2
1	$(1)^2 - 2$	−1
2	$(2)^2 - 2$	2
3	$(3)^2 - 2$	7

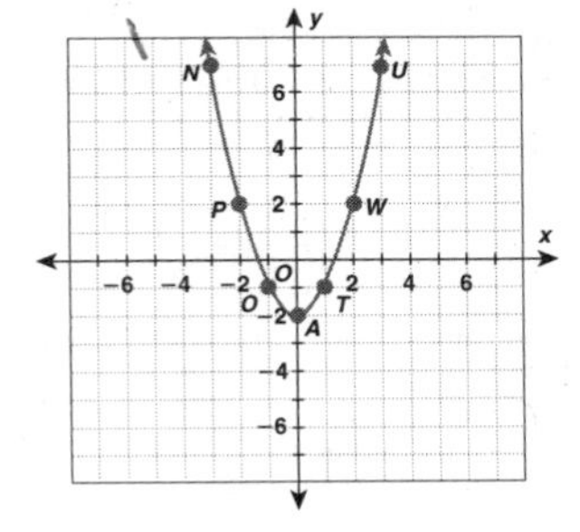

How does a dog stop a VCR?

By pressing the P (−2, 2) A (0, −2) W (2, 2) S B U (3, 7) T (1, −1) T O (−1, −1) N (−3, 7)

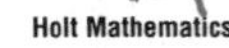

LESSON 13-7

Practice A

Inverse Variation

Complete the table for each inverse variation.

1. $xy = 60$

x	y
2	30
3	20
4	15
5	12

2. $xy = 1$

x	y
−8	$-\frac{1}{8}$
−2	$-\frac{1}{2}$
$\frac{1}{4}$	4
3	$\frac{1}{3}$

Tell whether each relationship is an inverse variation.

3.

x	y
6	4
3	8
2.5	9.6
−12	−2

The relationship is an inverse variation.

4.

x	y
9	8
3	24
0.25	288
0.06	120

The relationship is not an inverse variation.

5. Complete the table and graph the inverse variation function.

x	y
2	6
3	4
−8	−1.5
−4	−3
−2	−6

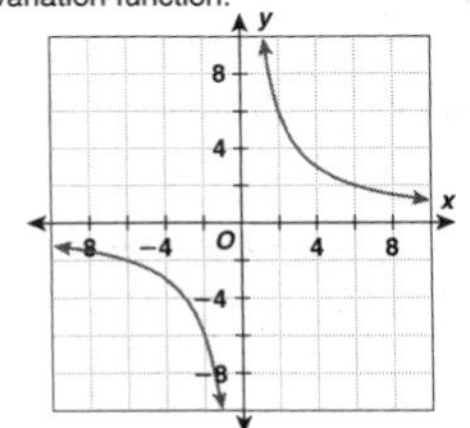

6. If x varies inversely as y and $x = 6$ when $y = 7$, what is the constant of variation? 42

LESSON 13-7

Practice B

Inverse Variation

Tell whether each relationship is an inverse variation.

1. The table shows the length and width of certain rectangles.

Length	6	8	12	16	24
Width	8	6	4	3	2

The relationship is an inverse variation.

2. The table shows the number of days needed to paint a house for the size of the work crew.

Crew Size	2	3	4	5	6
Days of Painting	21	14	10.5	8.5	7

The relationship is not an inverse variation.

3. The table shows the time spent traveling at different speeds.

Hours	5	6	8	9	12
mi/h	72	60	45	40	30

The relationship is an inverse variation.

Graph each inverse variation function.

4. $f(x) = \frac{4}{x}$

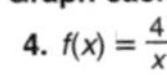

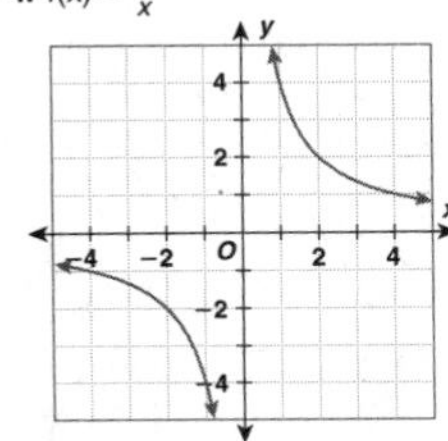

5. $f(x) = \frac{5}{x}$

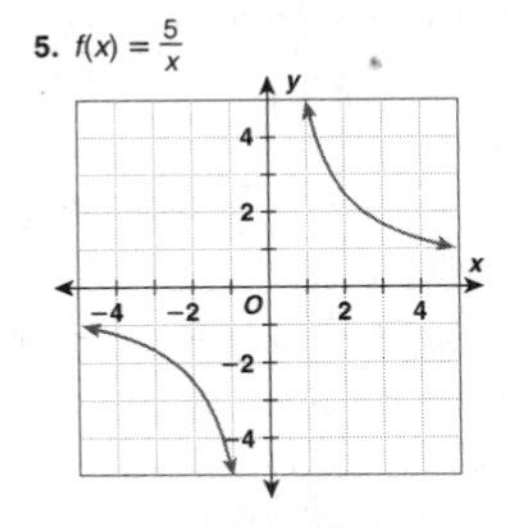

6. Amperes (abbreviated amp) measure the strength of electric current. An ohm is the unit of electrical resistance. In an electric circuit, the current varies inversely as the resistance. If the current is 24 amps when the resistance is 20 ohms, find the inverse variation function and use it to find the resistance in ohms when the current is 40 amps. $y = \frac{480}{x}$; 12 ohms

LESSON 13-7

Practice C

Inverse Variation

Find the inverse variation function, given that x and y vary inversely.

1. y is 4 when x is 22 $y = \frac{88}{x}$
2. y is 16 when x is 4 $y = \frac{64}{x}$
3. y is −12 when x is 18 $y = \frac{-216}{x}$
4. y is −15 when x is −50 $y = \frac{750}{x}$
5. y is 0.5 when x is 37 $y = \frac{18.5}{x}$
6. y is $\frac{1}{3}$ when x is 2115 $y = \frac{705}{x}$
7. If y varies inversely with x and if $y = -2$ when $x = -12$, find the constant of variation. 24
8. If y varies inversely with x and if $y = -\frac{2}{3}$ when $x = 48$, find the constant of variation. −32

For exercises 9–14, assume that y varies inversely as x.

9. If $y = -6$ when $x = -2$, find y when $x = 5$. 2.4
10. If $y = 200$ when $x = -\frac{1}{2}$, find y when $x = -2.5$. 40
11. If $y = 72.6$ when $x = 15$, find y when $x = 33$. 33
12. If $y = 15.5$ when $x = 3$, find y when $x = 5$. 9.3
13. If $y = -8$ when $x = 3.4$, find y when $x = 4$. −6.8
14. If $y = 49$ when $x = 14$, find y when $x = -7$. −98
15. Robert Boyle, a physicist and chemist is credited with what is now known as Boyle's law, $PV = k$, where P is the pressure, V is the volume measured in atmospheres and k is the constant of proportionality. Pressure acting on 30 m^3 of a gas is reduced from 2 to 1 atmospheres. What new volume does the gas occupy? 60 m^3
16. Suppose pressure acting on 30 m^3 of a gas is increased from 1 to 2 atmospheres. What new volume does the gas occupy? 15 m^3

LESSON **13-7**

Reteach

Inverse Variation

Two quantities **vary inversely** if their product is constant.

y varies inversely as *x*. $xy = k$ ← constant of variation

$y = \frac{k}{x}$ ← function of inverse variation

To determine if two data sets vary inversely, check for a constant product.

x	1	2	4	8	−1	−2	−4	−8
y	8	4	2	1	−8	−4	−2	−1

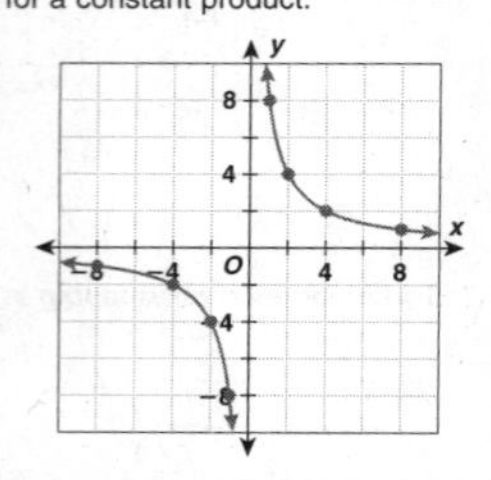

$xy = 1(8) = 2(4) = 4(2) = 8(1) = -1(-8)$
$= -2(-4) = -4(-2) = -8(-1) = 8$

So, *y* varies inversely as *x*.

The constant of variation is 8.

The function of inverse variation is $y = \frac{8}{x}$.

The graph of an inverse variation is two curves and is not defined for $x = 0$.

Tell whether the data shows inverse variation. If there is a constant product, identify it and write the function of inverse variation.

1.

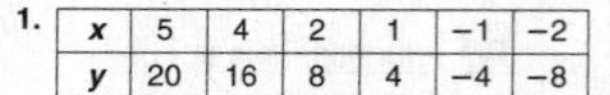

x	5	4	2	1	−1	−2
y	20	16	8	4	−4	−8

constant product? no

If yes, function. ______

2.

x	1	2	3	−3	−2
y	12	6	4	−4	−6

constant product? yes

If yes, function. $y = \frac{12}{x}$

Write the function of inverse variation. Then graph it.

3.

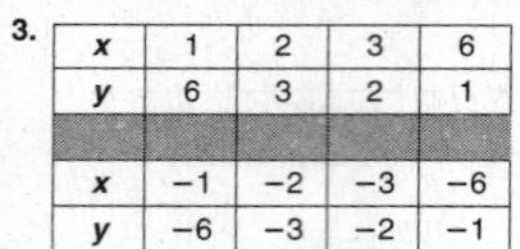

x	1	2	3	6
y	6	3	2	1
x	−1	−2	−3	−6
y	−6	−3	−2	−1

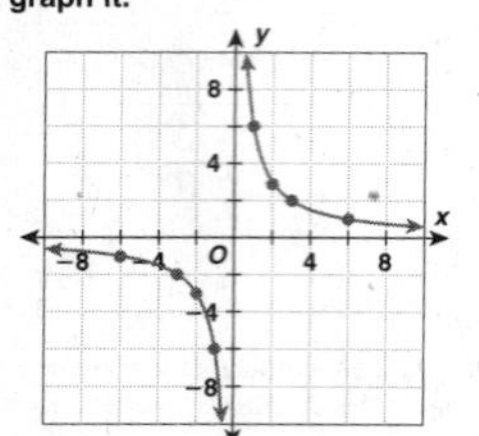

function: $y = \frac{6}{x}$

LESSON **13-7**

Challenge

When an Apple Fell on His Head!

The English physicist Sir Isaac Newton is said to have recognized the force of gravity as a result of having an apple fall from a tree under which he was seated. He later reasoned that any two objects attract one another gravitationally.

Newton's Law of Gravitation

The force of attraction *F* between two particles varies directly with the product of their masses, m_1 and m_2, and inversely as the square of the distance *r* between their centers.

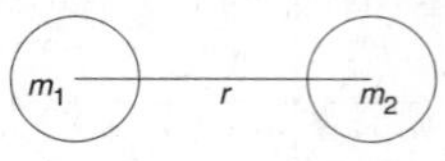

$F = \frac{G m_1 m_2}{r^2}$. *G* is the universal gravitational constant.

If the masses are expressed in kilograms, kg, the distance in meters, m, and the force in newtons, N, the value of the universal gravitational constant $G = 6.67 \times 10^{-11}$ N • m² / kg².

Find the gravitational force exerted by the Earth on a 1-kg mass on its surface.

mass of object, $m_1 = 1.0$ kg
mass of Earth, $m_2 = 6.0 \times 10^{24}$ kg
distance between = radius of Earth = 6.4×10^6 m

$$F = \frac{G m_1 m_2}{r^2} = \frac{6.67 \times 10^{-11}(1.0)(6.0 \times 10^{24})}{(6.4 \times 10^6)^2}$$

$$= \frac{(6.67)(1.0)(6.0) \times 10^{-11+24}}{(6.4)^2 \times 10^{6 \cdot 2}}$$

$$= \frac{40.02 \times 10^{13}}{40.96 \times 10^{12}} = 0.977 \times 10 = 9.8 \text{ N}$$

1. Find the gravitational force exerted by the moon on the Earth.
mass of moon = 7.3×10^{22} kg, Earth-moon distance = 3.8×10^7 m

$$F = \frac{G m_1 m_2}{r^2} = \frac{6.67 \times 10^{-11}(7.3 \times 10^{22})(6.0 \times 10^{24})}{(3.8 \times 10^7)^2}$$

$$= \frac{(6.67)(7.3)(6.0) \times 10^{-11+22+24}}{(3.8)^2 \times 10^{7 \cdot 2}}$$

$F = 2.0 \times 10^{22}$ N

2. Find the gravitational force exerted by the Earth on the moon.

$F =$ 2.0×10^{22} N, same as that of moon on Earth

LESSON **13-7**

Problem Solving

Inverse Variation

For a given focal length of a camera, the f-stop varies inversely with the diameter of the lens. The table below gives the f-stop and diameter data for a focal length of 400 mm. Round to the nearest hundredth.

f-stop	diameter (mm)
1	400
2	200
4	100
8	50
16	25
32	12.5

1. Use the table to write an inverse variation function.
$f(d) = \frac{400}{d}$

2. What is the diameter of a lens with an f-stop of 1.4?
285.71 mm

3. What is the diameter of a lens with an f-stop of 11?
36.36 mm

4. What is the diameter of a lens with an f-stop of 22?
18.18 mm

The inverse square law of radiation says that the intensity of illumination varies inversely with the square of the distance to the light source.

5. Using the inverse square law of radiation, if you halve the distance between yourself and a fire, by how much will you increase the heat you feel?
A 2
(B) 4
C 8
D 16

6. Using the inverse square law of radiation, if you double the distance between a radio and the transmitter, how will it affect the signal intensity?
(F) $\frac{1}{4}$ as strong
G $\frac{1}{2}$ as strong
H twice as strong
J 4 times stronger

7. Using the inverse square law of radiation, if you increase the distance between yourself and a light by 4 times, how will it affect the light's intensity?
(A) $\frac{1}{16}$ as strong
B $\frac{1}{4}$ as strong
C $\frac{1}{2}$ as strong
D twice as strong

8. Using the inverse square law of radiation if you move 3 times closer to a fire, how much more intense will the fire feel?
F $\frac{1}{3}$ as strong
G 3 times stronger
(H) 9 times stronger
J 27 times stronger

LESSON **13-7**

Reading Strategies

Use a Context

Jason drew different sketches for a dog pen with an area of 24 square units. Notice that as the length of the pen increases, the width of the pen decreases.

Length	Width	Area
2	12	24
3	8	24
4	6	24

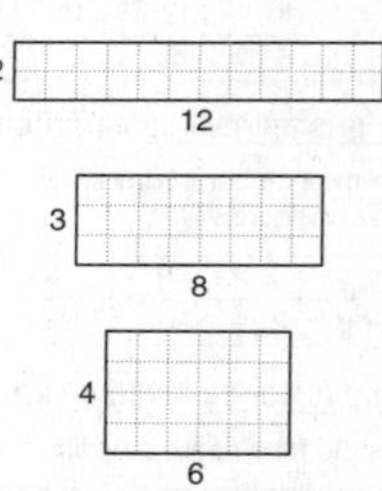

As one variable increases (length), the other variable decreases (width). The product of the variables (length and width) remains the same. (The area of each pen is always 24 square units.) This relationship is called an **inverse variation.**

Use the information above to answer each question.

1. Which dimension of the pen increases?
the length

2. Which dimension of the pen decreases?
the width

3. What stayed the same when the dimensions of the pen changed?
the area of the pen

4. What was the area of each pen?
24 square units

5. What name is given to the relationship between a variable that increases and a variable that decreases with a constant product?
inverse variation

LESSON **13-7**

Puzzles, Twisters and Teasers

Bounce This Around!

Fill in the blanks to complete the chart. Each answer has a corresponding letter.

Use the letters to solve the riddle.

Hint: $y = \frac{180}{x}$

Barn Construction Days and Size of Work Crew										
Crew Size	2	3	5	10	15	20	30	36	45	90
Number of Days	90	60	36	18	12	9	6	5	4	2
	E	T	N	R	I	A	L	M	O	P

What lived in prehistoric times, had very sharp teeth, and went boing-boing?

A T-Rex on a T R A M P O L I N E

T	R	A	M	P	O	L	I	N	E
3	10	20	5	90	4	6	12	36	2